EXPLOSIVES FOR
NORTH AMERICAN ENGINEERS

Other books published within the Series on Rock and Soil Mechanics

Editor-in-Chief
Professor Dr.-Ing. H. Wöhlbier

Series on Rock and Soil Mechanics
Vol. 3 (1978/79) No. 4

EXPLOSIVES FOR NORTH AMERICAN ENGINEERS

Second Edition

by

Cedric E. GREGORY
BE, BA *(Adel)*, BEcon, ME, PhD *(Qld)*
Chartered Engineer
Registered Professional Engineer
Professor Emeritus of Mining Engineering
in the University of Idaho, USA

TRANS TECH PUBLICATIONS

Distributed in North America by
TRANS TECH PUBLICATIONS
16 Bearskin Neck, Rockport, MA 01966, USA

and world-wide by
TRANS TECH S. A.
CH-4711 Aedermannsdorf, Switzerland

Copyright © 1979 by
Trans Tech Publications
D-3392 Clausthal, Germany

ISBN 0-87849-025-6

Printed in the United States of America

PRINTED AND BOUND IN THE UNITED STATES OF AMERICA

To my late colleague and mentor:
Professor Frank Thomas Matthews White,
whose distinctive talents
I have always admired

Acknowledgement of Photographs

Grateful acknowledgement is also made to the following who have given permission to reproduce photographs:

Austin Powder Co. (Figs.: 28, 35, 48)
Canadian Industries Ltd. (Figs.: 1, 4, 7—10, 12—16, 18, 19, 21, 24, 25, 32—34, 36)
Du Pont de Nemours & Co. (Figs.: 2, 3)
Hercules Inc. (Figs.: 5, 6, 23, 27)
Jet Research Center, Inc. (Figs.: 65, 66)
Vibration Measurement Engineers, Inc. (Fig.: 20)

Preface to First Edition

North American professional engineers at graduation have had little opportunity to embrace much of the rudimentary lore and practice of explosives. Many of them gain little more knowledge throughout their professional careers. Yet, to my mind, the intelligent use of explosives represents an important tool in many fields of engineering endeavor.

Although a sound knowledge of explosives principles is basic to the mining engineer, and very important to the civil engineer, I find, as a university teacher (in three American Universities), that the competition for lecture hours in our already congested curricula precludes the opportunity to cover the ground, even superficially, in regular classes.

It thus becomes necessary to deflect too large a part of the work to private reading by students. This again is a pious hope in view of the aforesaid crammed curricula and the meagre hours available for reading. But it would be significantly easier if there were available some publication dealing with the elementary principles and practices of those particular explosives marketed in North America and used under American and Canadian conditions.

It is too difficult and confusing for students in the time available to sift and screen separate disconnected items of trade literature to gain a firm conception of modern explosives practice in terms of the particular brands of explosives materials available; and the trade hand books available are too voluminous and confusing, quite unsuitable for teaching purposes, and restricted to the particular manufacturer's range of products.

Hence, I have been inspired to write this "primer" (pronounced with a short "i") or elementary monograph on the basic principles, types available, and methods applicable to the North American scene.

I make no claim to have covered the subject exhaustively either in breadth or depth. On the other hand, I have had great difficulty in confining the material to basic principles and generalities.

Moreover, I admit that I am out of fashion in this presentday world in aiming to prepare a textbook that is both readable and understandable: by presenting the subject matter clearly and omitting the more sophisticated though worthy analytical material. But in this way the book should also be of substantial benefit to operators in the field, and to their trainees at various levels. Otherwise, for those who would wish to delve deeper, an extensive reference list is provided at the end of each chapter.

As a necessary aid in clarification, I have condensed and classified the bewildering multitudinous array of explosives products available (some quite anachronistic and obsolete), along with their trade names. Naturally, even the trade names themselves may become outdated in time; this is one of the perils of publishing "useful" material.

This book is written in terms of Canadian spelling; however, as a concession, the words ending in -*our* have been translated into American -*or* forms.

The title is perhaps a misnomer, unless one is prepared to add *English text*. Apologies are therefore offered to my French-, Dutch-, Spanish-speaking friends in Canada, Mexico, Central America and the Caribbean zone. The preparation of a Latin-American edition in Spanish and Portuguese is presently under consideration.

A glossary is provided in case certain technological terms are not well understood by the reader.

Although I have drawn largely upon my own observations and experience over many years as a miner, as a mining engineer, as a mine manager, and as a mining professor, I desire to acknowledge my indebtedness to Nobel (Australasia) Pty. Ltd., upon whose technical material the Australasian editions (and a good deal of this edition, where applicable) have been based.

My personal thanks are also due to Mr. Harry Hampton Jr., Secretary of the Institute of Makers of Explosives, and to his Technical Committee, who have responded so magnificently in providing trade information from constituent members; to Mr. W. B. Morrey of Canadian Industries Ltd., to Mr. B. P. McHugh, Assistant Chief Inspector of Explosives of the Canadian Department of Energy, Mines and Resources, who reviewed the manuscript and proposed many helpful improvements; to Mr. Clyde W. Eilo (Hercules Incorporated) and to Dr. J. J. Yancik (U.S. Bureau of Mines) for reviewing my rough draft of Chapters 3 and 4 respectively; to Mr. Richard A. Dick (U.S. Bureau of Mines), upon whose excellent publications I have heavily drawn; to the National Fire Protection Association for the use of their material; to Austin Powder Company, Canadian Industries Ltd., E. I. DuPont de Nemours & Co. Inc., Hercules Incorporated, Jet Research Center Inc., and Vibration Measurement Engineers Inc., for provision of photographic and other material, not otherwise acknowledged; to Mr. Albert E. Teller, President of the International Society of Explosives Specialists, and to many others for their kind assistance and support.

Finally, I am indebted to the University of Queensland Press for permission to use certain material and diagrams appearing in my second Australasian edition.

C. E. Gregory

Preface to Second Edition

This new edition provides an opportunity to revise the original, to bring it up to date, and to give due emphasis to new techniques that have developed over the last five years. Unfortunately, I have been unable to predict the final outcome of new regulations currently under consideration by the U. S. Department of Transportation. However, I have dropped the term "Nitrocarbonitrate (NCN)" because it is meaningless, if not misleading.

I have always held that loose expressions and confusion in usage of terms are dangerous in this safety-conscious industry, especially among new trainees and neophytes generally. For this reason, I have made a conscious effort to crystallize some of these terms in the hope that manufacturers will be gracious enough to follow my example. For instance, I now use the term:

- **Detonating cord,** because 'detonating fuse' is liable to be con-fused with 'safety fuse'.

- **Boosters,** for the cast explosive units as supplied by the factories. They should not be called 'primers'. They do not become primers until armed with a detonator.

- **Slim boosters** (for want of a better name), to distinguish these slip-on sheath type boosters from ordinary boosters.

- **Boostering charges,** for the free-flowing explosive material used to supplement cartridged charges in a quarry hole.

- **Minimum initiator,** to replace the term 'minimum booster'.

- **Capped fuse,** instead of 'fuse primer', which is an abomination.

It is recognized that the NFPA definition of blasting agent, based upon cap-sensitivity, has been a great advantage to the explosives industry. This definition seems to have been justified over about twenty years of experience, except that recent rapid developments in slurries have produced a grey area in the definition. To overcome this problem, it appears logical to include cap-insensitive slurries under the heading of blasting agents (SBA); and to regard cap-sensitive slurries (or those containing an explosive ingredient) as slurry explosives (SE). I have therefore taken the liberty to make this distinction in the book. Accordingly, slurry explosives are now dealt with, among other high explosives, in Chapter 3; and slurry blasting agents, as usual, in Chapter 4. There seems no other practical choice than to split slurries in this way, even though other properties of slurries are uniquely similar in character.

I had planned to switch to *Système Internationale* (SI) units of weights and measures in this edition. Although Canada has moved in this direction, there has been no evident similar approach in the United States. Accordingly, I have been obliged to retain the use of English units.

In my classes over recent years I have been encouraged by the warm response to this book, not only by mining students, but also by civil engineering students. Otherwise, it is sad to note that so many civil engineering curricula of universities in the United States are devoid of courses in rock blasting, quarrying or tunnelling — subjects so vital to the civil engineering contractor.

As before, I am indebted to the Institute of Makers of Explosives and to their constituent members for providing trade literature. Where any of these members are not represented in the text, it is perhaps for lack of a sufficient postal address. They are encouraged to contact me for future reference.

Edificio Phoenix del Mar 411   C. E. Gregory
Los Arenales del Sol
Alicante, Spain

CONTENTS

CHAPTER 4   Blasting Agents

CHAPTER 5   Initiating Devices

CHAPTER 6   Firing with Conventional Safety Fuse

CHAPTER 7   Electric Shotfiring

CHAPTER 8   Shotfiring with Detonating Cord

CHAPTER 9   Sequential Firing

CHAPTER 10   Special Forms of Explosives

CHAPTER 11   Practical Usage of Explosives

PART II   BREAKING GROUND WITH EXPLOSIVES

CHAPTER 12   Introduction

CHAPTER 13   Blasting Theory

CHAPTER 14   Special Techniques

CHAPTER 15   Cut Holes

CHAPTER 16 Quarry and Open Cut Blasting Practice

CHAPTER 17 General Industrial Applications

CHAPTER 18 Blasting in Tunnels and Underground Development Headings

CHAPTER 19 Blasting in Stoping Operations

CHAPTER 20 Blasting in Coal Mines

PART I

EXPLOSIVES AND ACCESSORIES

ABBREVIATIONS

AWG	American Wire Gauge
FGAN	Fertilizer grade ammonium nitrate
ft	foot, feet
ft³	cubic foot (feet)
ft³/lb	cubic feet per pound
ft/sec	feet per second
ft³/ton	cubic feet per ton
gm	gram(s)
gm/cc	grams per cubic centimetre
gr/ft	grains per foot
in	inch(es)
lb	pound(s)
lb/ft³	pounds per cubic foot
lb/in²	pounds per square inch
lb/ton	pounds per ton
min	minute(s)
mm	millimetre(s)
msec	millisecond(s)
oz	ounce(s)
oz/ft²	ounces per square foot
oz/yd³	ounces per cubic yard
sec/ft	seconds per foot
USSS	United States Standard Sieve

CHAPTER 1

Introduction

Definition

The most generally accepted definition of an explosive is expressed in the following terms:

> A solid or liquid substance or mixture of substances which, on the application of a suitable stimulus to a small portion of the mass, is converted *in a very short interval of time* into other more stable substances, largely or entirely gaseous, with the development of heat and high pressure.

The explosive material derives its disrupting force, when confined in a borehole in a rock mass, mainly from the strain energy (SE) creating radial cracking in the rock, and from the "bubble energy" (BE) representing the high pressure gases in expanding these cracks to the point of brittle fracture (see Chapter 2). These hot gases occupy many times the original volume and thereby exert high pressure on the surrounding rock.

The stimulus in the above definition may be provided accidentally by friction, impact, or heat; but under controlled conditions by the shock wave produced by a detonator incorporated in the explosive charge (see Chapter 5).

Historical

Gunpowder, or blasting powder (BP), was first used in the thirteenth century. Its first recorded use in mines was in Slovakia in 1627. Blasting powder was introduced into Cornwall in 1689, but its use did not become general until a safe method of ignition (by safety fuse) was invented by WILLIAM BICKFORD in 1831.

Nitroglycerine (NG) and nitrocellulose (NC) were developed in 1846, but not used as explosives until NOBEL introduced the fulminate detonator in 1867. In the same year NOBEL made nitroglycerine safe to handle by absorbing it in kieselguhr, forming a plastic mass containing 75 per cent of NG. This was known as "Guhr Dynamite".

In 1875, NOBEL followed the invention of dynamite by that of blasting gelatine (BG). This was a gelatinous mixture, formed by incorporating 92 per cent of NG with 8 per cent of NC. It is still one of the most powerful of the industrial explosives.

This important development was followed in 1879 by his invention of lower strength explosives by the admixture of sodium nitrate (SN) and other materials to NG and NC. A wide range of explosives based upon these substances has since been developed.

Other important developments include (a) low-freezing types of nitroglycerine explosives; (b) permitted (permissible) explosives for use in underground coal mines; (c) electric instantaneous and delay detonators; (d) detonating cord; (e) liquid oxygen explosives; (f) shaped charges (first propounded by MUNROE over eighty years ago); and (g) binary (high energy) explosives.

In 1935, a canned variety of factory-mixed blasting agents, mainly consisting of ammonium nitrate in varying densities, was introduced in the United States; later, the development of "Akremite" in 1954—1955 gave an impetus to the "on-site" mixing of various types of blasting agents. Since then, there has been a phenomenal growth in the usage of this family of ammonium nitrate carbonaceous mixtures (generally known by the most popular type, AN/FO), giving rise to the recent development of metallized slurries.

Experiments conducted in Nevada and New Mexico since 1957 by the Project Plowshare team have successfully demonstrated the applicability of underground nuclear blasting for peaceful purposes.

Applications

Explosives are now employed in the following ways:

1. **Military purposes:**

For propellants in gun ammunition.

As a destructive agent in bombs, torpedoes, and grenades.

For demolition purposes.

For pseudo-civil purposes such as excavating, quarrying, road-making, etc.

2. **Industrial purposes:**

(a) *Mining*

By far the largest consumption of industrial explosives is involved in mining for the following specific applications:

For excavation of shafts, tunnels, headings, galleries, and general mine workings.

For recovery of minerals from mines and collieries.

For winning of non-metallic material such as limestone, ironstone, road metal, etc., from quarries.

For geophysical shotfiring (seismic blasting).

(b) *Civil Engineering*

For excavation of railway, highway and hydro-electric diversion tunnels.

For building and machinery foundations.

For road and railway formations in hilly country.

For general earthworks.

For submarine harbor development.

(c) *Agriculture*

For well-sinking.

For excavating irrigation and drainage ditches.

For removal of tree stumps.

For sinking post-holes for fencing.

For boulder blasting.

For loosening subsoil in orchard development.

(d) *Mechanical-electrical*

For preparing machinery foundations.

For demolishing or adjusting foundations in an existing power station.

For breaking up heavy machinery in an existing power station.

For blasting transmission line pole-holes.

The use of explosives has contributed largely to our present high standard of living by minimizing the arduous nature of rock excavation work and by increasing the productivity of such operations.

Indeed, many operations now regularly carried out with ease would not be possible without the use of industrial explosives.

Furthermore, certain major classes of excavation work (of extreme importance in national development programmes), now considered economically impracticable with the use of conventional explosives, may possibly be effected in the future by nuclear blasting.

Manufacture

Gunpowder was first made in North America in 1675 near Boston, MA for hunting purposes; and the first reported manufacture of dynamite under NOBEL's patents was in 1868 by the Giant Powder Company, which subsequently became Atlas Chemical Industries, Inc.

In the United States, industrial explosives and/or components are currently being manufactured by the following major suppliers. For convenience, they will be referred to later in this textbook by the short name appearing in brackets.

Apache Powder Company*	(Apache)
Atlas Powder Company*	(Atlas)
Austin Powder Company*	(Austin)
Coast Fuse*	(Coast)
E. I. DuPont de Nemours & Co., Inc.*	(DuPont)
Energy Sciences & Consultants, Inc.*	(ESC)
Ensign Bickford Company*	(Ensign)
Gearhart-Owen Industries, Inc.*	(Goex)
Gulf Oil Chemicals Company	(Gulf)
Hercules Incorporated*	(Hercules)
Independent Explosives Co. of Pennsylvania	(Independent)
IRECO Chemicals*	(IRECO)

Monsanto Company*   (Monsanto)

Phillips Petroleum Company   (Phillips)

Sierra Chemical Company*   (Sierra)

Trace X Chemical*   (Trace)

Trojan Division of IMC Chemical Group, Inc.*   (Trojan)

The names of firms appearing above with an asterisk (*) are members of the Institute of Makers of Explosives (of 420 Lexington Avenue, New York, NY. 10017), a trade association formed in 1913 to promote better safety practices within the industry—hereinafter cited as the IME .

In Canada, two of the manufacturers are:

Canadian Industries Ltd.   (CIL)

DuPont of Canada Co. Ltd.   (DuPont/Can)

There may be others in both countries.

All production operations are closely controlled, and rigorous tests are made at each stage of manufacture. The results of research into newer methods, better performances, greater safety, and more durable packaging media are constantly being applied.

Apart from the manufacturers of explosive materials listed above, others supply a wide range of accessories to service the explosives industry. Among these are:

Amerind MacKissic, Inc.   Borehole dewatering units
                                        Bulk transport trucks
                                        Cartridge packers
                                        Mixing and blending equipment

Baughman Manufacturing Co.   Bulk transport trucks

Brentwood Plastic Films, Inc.   Lay-flat plastic tubing

Coleman Cable and Wire Co.   Blasting wire

Dallas Instruments, Inc.   Blast monitors
                                        Seismographs

Dearborn Wire and Cable Co.	Blasting wire
Fidelity Electric Co., Inc.	Blasting machines
Hydrophilic Industries, Inc.	Drill angle indicators
Jamer Corporation	Plastic collar plugs
Jarvis Clark Co. Ltd.	Mobiloders
Jet Research Center, Inc.	Shaped charge units
Research Energy of Ohio.	Sequential blasting machines
Seminole Products Co.	Blasting wire
Swanson Engineering Co.	Blasthole dewatering systems
Thyssen.	Trabant gel-stemming
USS Agri-Chemicals.	Blasthole dewatering systems
VME-Nitro Consult, Inc.	Blasting machines Bulk transport trucks Lightning detectors Mixing units Ohmmeters Pneumatic cartridge chargers Pneumatic charging machines Seismographs Sound testing service Tamping machines Velocity recorders.

Technical Service

Most manufacturers maintain a field technical sales service for their customers and industrial users generally. These representatives perform an excellent service and are readily available to help with technical recommendations, trouble-shooting, sales problems and the like. However, the mining, quarrying and construction industries should not lose sight of the fact that more than one-half of the companies engaged in these industries could realize better costs if their own engineers were better trained in the use of explosives.

A number of consulting companies offer their services to the explosives industry in the United States and Canada. Some of these specialize in demolition work and in ground vibration analysis.

In addition, a number of associations with a central interest in the explosives industry have been founded in North America. Among these are:

> American Blasting Association
> Institute of Makers of Explosives
> International Society of Explosives Specialists
> Society of Explosives Engineers.

Safety Training

In the interests of safety, DuPont conducts a 3-day Safety Training Course, developed to meet the needs of all personnel in the explosives industry. These courses are scheduled at different centres in the United States throughout the year. The course is of a practical nature and no educational pre-requisites are necessary.

CHAPTER 2

Rationale of Explosives Usage

Characteristics

In order to fulfil the requirements of everyday modern industrial usage, an explosive should have the following characteristics:

1. It should contain a substance, or mixture of substances, which is sufficiently insensitive to be safe under all conditions of handling and storage, but is sensitive enough to initiate readily when required.

2. Upon initiation it must rapidly undergo chemical change, yielding gaseous products whose volume under normal pressure and at the high temperature resulting from the exothermic reaction is much greater than that of the original substance.

3. The reaction must be exothermic in order to increase the pressure.

4. The substance should be simple and inexpensive to produce from readily available raw materials.

5. It should have:

    (a) Adequate strength or power for the purpose required.

    (b) A high inherent velocity of detonation (except where excessive shattering is to be avoided).

    (c) A density suited to its particular application.

    (d) Good water-resistance (or be specially wrapped and dipped to promote water-resistance).

    (e) Good fume characteristics.

(f) No propensity to freeze or dissociate (e.g., by exudation of nitro-glycerine) at the working temperature.

(g) Suitable physical characteristics, depending upon the inclination of the borehole.

(h) Good storage qualities.

Properties

The properties discussed below are weight strength, cartridge strength, velocity of detonation (VOD), sensitivity, density, detonation pressure, water-resistance, and fume characteristics. These properties vary with different types and formulations of explosives as prepared by manufacturers.

The **Strength** of an explosive is related to its energy content and is therefore regarded as a measure of its ability to do useful work. However, it is not a good yardstick.

In the past, the strength of a particular dynamite type explosive was determined by comparing its capacity to deflect a ballistic mortar with that of Blasting Gelatine or TNT (see Glossary). In the trade, two ratings of strength are used, both related to the nitroglycerine (NG) content of various grades of "straight dynamites" (see later) as a standard.

For instance, equivalent **Weight Strengths** (lb for lb) of ammonia gelatines and ammonia dynamites are given the same percentage ratings (gradings) as those corresponding to the NG content of a straight dynamite. A 50 % ammonia dynamite has the same strength as a 50 % straight dynamite even though it contains much less than 50 % NG. Actually, it contains ammonium nitrate (AN) in addition.

In the same way, the **Cartridge Strength** of a 50 % ammonia dynamite has the same cartridge strength as that of a 50 % straight dynamite. Since the specific gravity of straight dynamite is 1.4, the weight strength and cartridge strength of a given explosive are equal when its density is also 1.4; otherwise the relationship depends upon its density (see later).

In any case, DICK[1] has pointed out that the field performances of two different dynamites of the same weight strength may differ because of a difference in VOD. This shows that "strength" is not a useful parameter for the rating of explosive performance.

Some manufacturers rate their explosives on weight strength and others on a cartridge strength basis. Some do not even rate them at all in their literature, but refer to them under a trade name. Naturally, all manufacturers use trade names for their products, but many of these have little or no logical significance or consistency.

The strength of blasting agents is expressed on a different basis. CIL quotes the weight and bulk strengths of their blasting agents relative to that of "Standard AN/FO", which has an assigned RWS and RBS value of 100. Standard AN/FO is defined as a mixture of prilled AN having a 1 % inert coating and 5.7 % diesel fuel, resulting in a product whose density is 0.84 and which has an oxygen balance of 0.5 % deficient.

The **Velocity of Detonation** (VOD) refers to the speed at which a one-dimensional detonation wave (see later) representing a shock front, travels through a charge of explosives confined in a borehole.

The VOD of an explosive depends upon various other factors, such as the density, the constituents, their particle size, the diameter of the hole (or charge) and the degree of confinement.

Some manufacturers test under unconfined conditions, in which case the unconfined value is only about 75 per cent of the confined VOD. VOD values are best expressed in ft per second when tested confined in a pipe of a specified diameter at 40 °F. In general, the VOD increases with the diameter, but not linearly. With commercial explosives, the VOD ranges from 5000 to 25,000 ft per second.

Two areas of **Sensitivity** are recognized in respect of explosive properties. One of these is termed **Hazard Sensitivity** and relates to the "relative hazards involved in the formulation, handling, storing, and transportation of the explosive."[2]

The other is generally referred to as **Performance Sensitivity** since it measures three different aspects of reliability of performance of explosives under certain specified conditions.

One of these is termed **Initiation Sensitivity** — the ease with which an explosive charge can be detonated. Most cartridged high explosive charges (dynamites and gelatines) can be detonated when primed with a single No. 6 detonator (see Glossary). It is in the category of (insensitive) blasting agents that initiation sensitivity becomes critical. The threshold value of initiation sensitivity for a particular charge (under certain specified conditions) is expressed as the **minimum initiator** (or the minimum number of No. 6 caps) needed to detonate the charge.

The second aspect of performance sensitivity is usually distinguished by the name of **Propagation Sensitivity,** which is a measure of the ability of an explosive to support or reinforce a detonation that has already been created.[3, 4] Since the tendency for the detonation wave to fail or fade increases as the hole diameter is reduced, it has become customary to measure propagation sensitivity by the (minimum) **critical diameter** of a charge at least six diameters long that will support the detonation process once created.[3, 4] At a point below this value, a detonation wave cannot be consistently propagated, and a misfire (see Glossary) may occur; or otherwise the reaction may be incomplete, resulting in a poor blast effect and significant fume.

The third aspect of performance sensitivity is known as **Gap Sensitivity,** determined by the comparative length of air gap or discontinuity between two cartridges in a single charge.[2] It also relates to the distance between two adjacent blastholes in cases where *sympathetic detonation* could be a problem, or in a certain type of ditch blasting (by the propagation method), where it is an advantage. This phenomenon is peculiar only to (sensitive) high explosives such as dynamites and gelatines.

The **Density** of a particular explosive is expressed either in terms of specific gravity, or by cartridge count. Since the standard case (box or carton) in the United States contains 50 lb weight of explosives, it is conventional to quote the cartridge count as the number of $1^{1}/_{4}$ x 8-inch cartridges in a case; but in Canada, a 25 kg standard case is now used.

Explosives presently available range in specific gravity from 0.6 to 1.7, and with cartridge counts from 232 to 83 in the United States.

With free-running blasting agents, it is the loading density that is significant: the weight of material per ft of charged borehole of a given diameter; or otherwise in grams per cc. This measure will depend upon the degree of compaction achieved when loading. With wet holes, it needs to be sufficient to displace the water.

The **Detonation Pressure** is a measure of the pressure in the detonation wave front. It is a function of the VOD and the density of an explosive, and is therefore a very important property, especially in hard rock. At the point where the detonation wave has just completed its passage through the column of explosive and the gases have expanded to fill the borehole and are available to commence their work, the borehole pressure is at its peak value.[5]

Water Resistance, for dry boreholes, is not required; but for prolonged exposure to water, as in wet holes, an explosive needs to have very good water resistance to avoid deterioration or loss of sensitivity. The more gelatinous types of explosive have the better resistance to water. These are especially suitable for under-water blasting.

The **Fume Characteristics** of an explosive are of extreme importance when used in confined spaces as in underground mines.

Ideally, industrial explosives should be designed to produce gaseous products such as nitrogen, carbon dioxide and water vapor. In practice, some toxic gases such as mixed oxides of nitrogen and carbon monoxide (known collectively as "fumes") may be formed. These are extremely dangerous to inhale in certain concentrations.[6]

The U.S. Bureau of Mines and the IME have together established a Fume Classification Standard based upon the volume of poisonous gases emitted by a $1^1/_4 \times 8$-inch cartridge when detonated under standardized conditions. Explosives may be classified under this standard as follows:

Class 1:   0 to 0.16 ft³ per cartridge

Class 2:   0.16 to 0.33 ft³ per cartridge

Class 3:   0.33 to 0.67 ft³ per cartridge

Only those explosives meeting the requirements of Class 1 may be used in underground mines. For coal mines, the volume of poisonous gases produced must not exceed 71 litres per 454 gm of permissible explosive (2.5 ft³/lb). In any event, good ventilation is a necessary requirement; and the need to disperse the fumes from a blasted heading before miners return to work is a vital precaution.

Explosives in North America are formulated with a low freezing point component to prevent **freezing** under the coldest weather conditions. Heated storage magazines are therefore unnecessary, even for slurries. Without this component, dissociation and exudation of NG and deterioration of the explosive would be likely to occur.

Testing of Explosives

A number of methods of testing or checking the various properties of explosives have been proposed or recognized in the past. The development of more sophisticated methods is constantly being applied as improved instruments and procedures become available. However, since testing is considered to be primarily a phase of the manufacturing process, details will not be offered in this book. A review of methods currently used by the U.S. Bureau of Mines has been prepared by MASON and AIKEN.[7]

Types of Explosives

The exothermic decomposition of an explosive produces exceedingly high gas pressures at the prevailing high temperature of the reaction.

Therefore the effectiveness of an explosive depends upon the amount of heat liberated.

Although the explosive energy is equi-directional, its rupturing effect is greatest in the path of least resistance.

If the explosive decomposition reaction moves through the charge faster than the speed of sound, it is termed *detonation*. If slower, then we have *deflagration*.

All high explosives and blasting agents detonate when properly initiated, whereas low explosives (black powder) deflagrate.

There are four main types of modern commercial explosive. One of these is the class of initiating explosives used only as a loading medium in detonators and detonating cord.

The selection of an explosive for a given engineering duty such as rock breakage may be made from the three other main types, viz.

(a) high explosives (TNT, dynamites and slurry explosives)

(b) low explosives (blasting powder), or

(c) blasting agents (dry types or slurries).

1. **Initiating explosives,** because of their high cost, are suitable only for *initiating* high explosive charges as detonating agents in small quantities. Upon initiation, they produce an intense shock capable of initiating a detonating wave in a charge of high explosives. These explosives are loaded as small pressed charges into copper or aluminium tubes to form detonators (q.v.), or as a core load in detonating cord.

2. **High explosives,** according to their composition, *detonate* at velocities of 5,000—25,000 feet per second and produce large volumes of gases and considerable heat at extremely high pressure. The performance of such

an explosive depends chiefly on the volume and temperature of the gases produced and on the velocity of detonation.

High explosives usually contain chemical components with a more or less unstable molecular structure, capable, on detonation, of molecular re-arrangement into more stable forms of potential energy. Thus, the compound is capable, within its own volume, of being converted into gases at very high temperature and pressure. The entire products of the explosion should be gaseous. (Solid products would tend to produce smoke.)

As an example, the chemical change that takes place when nitroglycerine explodes is shown by the following equation:

$$4C_3H_5(NO_3)_3 \rightarrow 12CO_2 + 10H_2O + 6N_2 + O_2$$

Similarly, the typical chemical reaction which takes place when a mixture of nitroglycerine and ammonium nitrate is exploded is:

$$2C_3H_5(NO_3)_3 + NH_4NO_3 \rightarrow 6CO_2 + 7H_2O + 4N_2 + O_2$$

It will be noted that in each case, at the temperature of the reaction, all the products are gaseous.

3. **Deflagrating or low explosives** were the earliest type developed. This type of explosion is really a rapid form of combustion in which the particles *burn* at their surfaces and expose more and more of the bulk until all has been consumed.

Typical of these low explosives is Blasting Powder or gunpowder which is a mixture of substances having an affinity for oxygen with others which are rich in oxygen. The common form is black blasting powder which is a mechanical mixture of potassium or sodium nitrate with sulphur and finely ground charcoal.

In action, they are characterized by a *push*, rather than a *blow*, or a *lift* rather than a *shatter.*

Propellants in ammunition are low explosives. The slow gradual development of pressure in the chamber of a gun allows the projectile to issue with a high muzzle velocity without shattering the gun barrel. Since about one-half the products of combustion are solids, much smoke is formed.

$$20NaNO_3 + 30C + 10S \rightarrow 6Na_2CO_3 + Na_2SO_4 + 3Na_2S_3 + 14CO_2 + 10CO + 10N_2$$

4. **Blasting agents** are mixtures "consisting of a fuel and an oxidizer, intended for blasting, not otherwise classified as an explosive and in which none of the ingredients [is] classified as an explosive, provided that the finished product, as mixed and packaged for use or shipment, cannot be detonated by means of a No. 8 test blasting cap when unconfined".[8]

They are classified as "oxidizing materials" under the U.S. Department of Transportation (DOT) regulations (see Chapter 22), and are thereby exempted from the restrictive regulations necessarily applying to the transport of high (cap-sensitive) explosives. However, these regulations are currently under review for transport of blasting agents in bulk. In Canada, they are classified as "Explosives Class 2", and transport and mixing procedures are more closely controlled.

A blasting agent consists primarily of inorganic nitrates and carbonaceous fuels, and may contain additional nonexplosive substances such as powdered aluminium or ferrosilicon.[1]

Blasting agents (of which the more generally known is AN/FO) are now extensively used for both surface and underground blasts. The relatively inactive AN, in a suitable physical form, can be sensitized by mixing with fuel oil (in the proportion of 94:6 by weight), or other carbonaceous fuel. It will detonate at a velocity of 10,000 to 14,000 feet per second (generally at a total rock breakage cost of about one-third to one-quarter of that of NG-based high explosives) according to the following general reaction:

$$3NH_4NO_3 + CH_2(X) \rightarrow 7H_2O + CO_2 + 3N_2$$

Apart from AN/FO, there are **slurry blasting agents** (SBA) which are of greater density, higher detonation velocity, and of very much greater water resistance. See Chapter 4 for further details.

Mechanics of Detonation

Under the heading of "Deflagration and Detonation", CLARK[9] states:
"Explosives either react slowly, or deflagrate, or they may detonate. A distinguishing characteristic of a high explosive is that it detonates when it is properly primed and an explosion is initiated in it. **Detonation** is the description of the process of the propagation of a shock wave through an explosive, which is accompanied by a chemical reaction that furnishes energy to maintain the explosion in a stable

manner. Black powder is a good illustration of deflagrating explosives while [high explosives] are representative of detonating explosives."

Although **high explosives** will burn or deflagrate if small amounts are thermally ignited, large masses (above the *critical mass* for any given explosive) will deflagrate progressively more fiercely until detonation takes place (without the actual presence of a detonator).

However, for proper and effective control, a charge of high explosive confined in a borehole is usually *initiated* by some type of **initiating explosive,** in the form of a **detonator** (see Chapter 5). The detonator is usually incorporated in a cartridge of high explosive to form a **primer.** The detonator itself is thermally ignited, if plain, by **safety fuse** (see Chapter 6); or if electric, by an electrical impulse.

The thermal ignition of the initiating explosive in the detonator is referred to as **detonation.** This is a reactive shock, caused by instantaneous exothermic decomposition of the initiating explosive, accompanied by rapid expansion of the reactive products, thereby producing a high pressure shock, or detonation, wave.

This shock wave proceeds along the high explosive charge with a *transient* velocity of detonation for some finite distance corresponding to a very minute interval of time until the main charge is initiated (by the high pressure and temperature of the gaseous products of the detonation reaction) at the CHAPMAN-JOUGET Plane[10] (see Glossary).

At this C-J Plane, the detonation wave is reinforced and propagated throughout the charge by the rapid exothermic decomposition of the main charge, at what is known as the *steady-state* velocity of detonation (usually a constant speed varying from 5,000 to 25,000 ft/sec, depending upon the performance parameters of the particular charge).

The **explosive coupling**[3, 11] (or **coupling ratio)** relates to the efficiency with which the chemical energy of the explosive is transferred to the enclosing rock of the borehole. With good coupling, there is little energy lost (e.g., through air space between the charge and the wall of the borehole).

Rock may be broken by an explosive charge according to the **Reflection Theory** propounded by HINO[12] and elaborated by DUVALL and ATCHISON as follows:[13]

"The detonation of the explosive charge creates a high gas pressure in the charge hole, which in turn generates a compressive strain pulse in the surrounding rock. This compressive strain pulse travels outward in all directions from the charge hole. Near the charge hole the amplitude of the strain pulse is sufficient to cause crushing of the rock. However, as the strain pulse travels outwards, its amplitude decays rapidly until no further crushing of the rock is possible. The compressive strain pulse continues to travel outward until it is reflected by a free surface. Upon reflection, the compressive strain pulse becomes a tensile strain pulse. As the strength of the rock in tension is much less than in compression, the reflected tensile pulse is able to break the rock in tension, progressing from the free surface back towards the shot point. In other words, the rock is pulled apart, not pushed apart. . . . The high gas pressure generated by the detonation of the charge produces the stress waves, but the expanding gases are not directly responsible for much of the fracturing that occurs."

Since most hard tenacious rocks are characterized by high compressive, lower shear and much lower tensile strength, it is of course logical to aim to break the rock in tension.

However, the phenomenon of spalling, as characterized by the reflection theory, is not now regarded as the only cause of rock failure. Several other factors contribute. The peak strain energy (SE) developed at the wall of the blast hole by high detonation pressure is designed to overcome the dynamic compressive strength of the rock. This compressive strain wave causes crushing and cracking around the blast hole. If a free face occurs within its zone of influence, a reflected tension wave will generate external spalling from the free face. The high borehole pressure (bubble energy, BE) developed by the expanding gases then extends the cracks to the point of flexural rupture of the rock, aided by the external spalling.[14]

According to HAGAN,[15] the relative values of strain and bubble energy are more reliable indicators of explosive effectiveness than strength data.[15] Strain energy is more effective in hard brittle rocks; and bubble energy in rocks of lower degrees of elasticity. Different explosive formulations provide varying SE:BE ratios.

A study of the respective contribution made in breaking rock (a) by the stress waves reflected from the free face; and (b) by the expanding gases has been conducted by CLARK and SALUJA,[16] who infer that the latter cause preponderates where lower velocity explosives are used.

However, no generally acceptable quantitative theory has yet been developed to explain the rock failure process.[17]

It should also be noted that, since lower velocity explosives are more effective in breaking rock material exhibiting significant plasticity (see Chapter 14), the relative elasticity of the rock to be broken is a contributing factor.

The above discussion dealing with the mechanics of detonation refers to high explosives and blasting agents.

On the other hand, **low explosives** may be thermally ignited directly by means of a safety fuse without the interposition of a detonator. This means that they burn or deflagrate, yielding gases and solids more slowly and at lower but sustained pressures, thereby developing a heaving action, more applicable to the blasting of soft plastic rock and coal (where lump size is to be retained, and where firedamp conditions allow).

Black powder is now used mainly in the blasting of dimension stone and slate, where a minimum amount of shattering can be tolerated. Otherwise, in finer gradings, it is used as a core load in the manufacture of safety fuse.

Comparison of Basic Prime Costs

Comparative prime costs of different explosive materials are given below, based upon information published by DICK,[18] and using bulk AN prills as a base range of 100—200.

Bulk AN prills	100—200
Bulk AN/FO	150—300
Bagged AN/FO	200—500
Cartridged AN/FO	300—600
Bulk or bagged AN slurries	400—1250
Dynamites	750—1600
Gelatine dynamites	850—1700

Nevertheless, when total costs per ton of ore or rock are considered, the differences may become subject to modification.

Explosives Consumption

Based on figures supplied for consumption in the United States in 1976, the following estimate of explosives usage is offered.[19, 20]

AN prills and AN/FO	82 %
AN slurries	10 %
Dynamites and gelatines	6 %
Permissible explosives	1 %

Figures available for 1976 show a total consumption in the United States of approximately 1.66 million tons of explosives, represented in the following categories of industry:

Coal mining	54 %
Metalliferous mining	15 %
Quarrying etc.	15 %
Construction work	10 %
All other purposes	6 %

Blasting agents accounted for 92 per cent of the total consumption.

References

1. R. A. DICK, "Factors in Selecting and Applying Commercial Explosives and Blasting Agents," *U.S.B.M., I.C.* 8405 (1968).
2. M. A. COOK, *The Science of Industrial Explosives* (IRECO Chemicals, Salt Lake City, 1974).
3. J. J. YANCIK, "ANFO Manual," (St. Louis: Monsanto Company, 1969).
4. J. J. YANCIK, R. F. BRUZEWSKI, and G. B. CLARK, "Some Detonation Properties of Ammonium Nitrate," Fifth Annual Symposium on Mining Research, *University of Missouri School of Mines and Metallurgy, Bulletin,* no. 98 (1960).
5. A. BAUER, "Current Drilling and Blasting Practices in Open Pit Mines on a World Basis," Paper presented at 1971 Mining Show of American Mining Congress (Las Vegas, Nevada, October 11-14, 1971).
6. B. D. ROSSI, ed. "Control of Noxious Gases in Blasting Works, and Methods of Testing Industrial Explosives," VZRYVNOE DELO. Collection no. 68/25 (Translated from the Russian).
7. C. M. MASON and E. G. AIKEN, "Methods for Evaluation Explosives and Hazardous Materials," *U.S. Bureau of Mines I.C.* 8541 (1972).
8. National Fire Protection Association "Manufacture, Storage, Transportation, and Use of Explosives and Blasting Agents 1973," Pamphlet no. 495, (1973): pp. 6-7.
9. G. B. CLARK, "Mathematics of Explosives Calculations," Fourth Annual Symposium on Mining Research, *University of Missouri School of Mines and Metallurgy, Bulletin,* no. 97 (1959).

10. C. H. JOHANSSON and P. A. PERSSON, *Detonics of High Explosives* (London: Academic Press, 1970).

11. T. C. ATCHISON, "The Effect of Coupling on Explosive Performance," *Quarterly of Colorado School of Mines*, vol. 56, no. 1 (1961).

12. K. HINO, *Theory and Practice of Blasting* (Asa, Japan: Nippon Kayaku Co., 1959).

13. W. I. DUVALL and T. C. ATCHISON, "Rock Breakage by Explosives," *U.S. Bureau of Mines, R.I.* 5356, (1957).

14. R. L. ASH, "Improving Productivity through Better Blasting Control," Paper presented to AIME Annual Meeting (New York, February, 1975).

15. T. N. HAGAN, "Rock Breaking by Explosives," *Australian Geomechanics Society*. National Symposium on Rock Fragmentation (Adelaide, February 26-28, 1973).

16. L. D. CLARK and S. S. SALUJA, "Blasting Mechanics," Trans. AIME, vol. 229 (March 1964).

17. G. D. JUST, "The Mechanics of Rock Breakage by Explosives," (University of Queensland, Rock Breaking Seminar, February 24, 1971).

18. R. A. DICK, "Current and Future Trends in Explosives and Blasting — Part I", *Pit and Quarry*, July 1971.

19. R. A. DICK, "The Impact of Blasting Agents and Slurries on Explosives Technology", *U.S. Bureau of Mines, I. C. 8560*, 1972.

20. Apparent Consumption of Industrial Explosives and Blasting Agents in the United States, 1976. Mineral Industry Survey. U.S. Department of the Interior, Bureau of Mines.

CHAPTER 3

High Explosives

Introduction

Most high explosives consist of mechanical mixtures of two or more explosive bases and other additives (see Table 1).[1]

In formulating explosive mixtures, it is necessary to achieve an *oxygen balance* in order (a) to maximize the explosive energy of the reaction by aiming at complete combustion, and (b) to avoid the presence of toxic gases in the combustion products.

This is generally achieved by using (a) combustibles/fuels (see Table 1) to react with the excess oxygen and thereby prevent the formation of toxic oxides of nitrogen; or (b) oxidizing agents (oxygen carriers) to promote full oxidation of the carbon to carbon dioxide, thereby inhibiting the formation of toxic carbon monoxide.

An absorbent (see Table 1) such as kieselguhr has been used (since 1867) to absorb liquid explosives such as nitroglycerine. Antacids promote stability in storage. Low freezing point compounds reduce the propensity of the mixture to freeze in cold weather. Gelatinizing agents promote water resistance (as for the gelatine dynamites). Flame depressants or coolants reduce the size, duration and temperature of the flame during the explosive reaction to avoid methane/coal dust explosions in coal mines (see Chapter 20).

In the category of high explosives, the chief explosives bases are nitroglycerine (NG), and ammonium nitrate (AN).

Other explosive bases listed in Table 1 are used predominantly as gelatinizing agents, low freezing compounds, initiating explosives (for use in detonators or detonating cord), black powder (the typical low explosive),

TABLE 1

Ingredients used in Explosives

Ingredient	Chemical formula	Function
Ethylene glycol dinitrate (nitroglycol)	$C_2H_4(NO_3)_2$	Explosive base; lowers freezing point
Nitrocellulose (NC) (guncotton)	$C_6H_7(NO_3)_3O_2$	Explosive base; gelatinizing agent
Nitroglycerine (NG)	$C_3H_5(NO_3)_3$	Explosive base
Nitrostarch (NS)		Explosive base; "nonheadache" explosives
Trinitrotoluene (TNT)	$C_7H_5N_3O_6$	Explosive base
Metallic powder	Al	Fuel-sensitizer
Black powder (BP)	$NaNO_3+C+S$	Explosive base; deflagrates
Pentaerythrioltetranitrate (PETN)	$C_5H_8N_4O_{12}$	Explosive base; caps, detonating cord
Lead azide	$Pb(N_3)_2$	Explosive base; used in blasting caps
Mercury fulminate	$Hg(ONC)_2$	Do.
Ammonium nitrate (AN)	NH_4NO_3	Explosive base; oxygen carrier
Liquid oxygen (LOX)	O_2	Oxygen carrier
Sodium nitrate (SN)	$NaNO_3$	Oxygen carrier; reduces freezing point
Potassium nitrate	KNO_3	Oxygen carrier
Ground coal	C	Combustible, or fuel
Charcoal	C	Do.
Paraffin	C_nH_{2n+2}	Do.
Sulphur	S	Do.
Fuel oil	$(CH_3)_2(CH_2)_n$	Do.
Wood pulp	$(C_6H_{10}O_5)_n$	Combustible; absorbent
Lampblack	C	Combustible
Kieselguhr	SiO_2	Absorbent; prevents caking
Chalk	$CaCO_3$	Antacid
Calcium carbonate	$CaCO_3$	Do.
Zinc oxide	ZnO	Do.
Sodium chloride	NaCl	Flame depressant (permissible explosives)

and trinitrotoluene (TNT) which has a limited application in the industrial explosive area.

We now have three different categories of high explosives, each of differing physical characteristics and chemical composition; but all having comparable values as explosives in blasting rock. All are ranked as "Explosives Class A or B" and must be transported, stored, handled and used in compliance with all applicable U.S. federal, state, and local laws and regulations. In Canada, they are ranked as "Explosives Class 2".

These three categories are (a) Trinitrotoluene (TNT), (b) Dynamites and gelatines, and (c) Slurry explosives. They are dealt with separately in some detail below.

Trinitrotoluene

Trinitrotoluene (TNT) is a high density explosive of excellent water resistance. Although of considerable importance in the field of military explosives, it is manufactured commercially in North America only by CIL, in the form of pellets, flakes or grains. It combines the important advantages of a high strength explosives base with that of low hazard sensitivity.

Pelleted TNT is produced as smooth pellets about one-eighth of an inch in diameter by CIL under the trade name of 'Nitropel'. They perform as well as 60 per cent ammonia gelatines, with a blasthole loading density of 0.94 grams/cc.

Individual pellets have a density of 1.5 and therefore they sink readily in water.

Nitropel is used in a variety of ways:

(a) As a supplement to rigid dynamite or blasting agent cartridges in large diameter quarry holes, to improve loading density and coupling ratio;

(b) as a sensitizer in the formulation of certain slurry explosives; and

(c) in seismic prospecting applications.

Nitropel can be poured direct into wet and dry quarry holes either as the main charge, or to fill the annular space around rigid cartridges as a supplement. In the latter case, it is primed by the surrounding charge. When used as a main charge, it calls for a high strength primer. When detonated, much fume is produced; therefore TNT cannot be used underground.

As a main charge, the loading densities achieved with TNT in holes of different diameters are given in the table below. These figures are based upon a pellet density of 0.94 grams/cc.

Hole Diameter	Loading Density	
(inches)	(lb/ft)	(kg/m)
2	1.3	1.9
$2^1/_2$	2.0	3.0
3	2.9	4.3
$3^1/_2$	3.9	5.8
4	5.1	7.3
$4^1/_2$	6.5	9.5
5	8.0	11.9
$5^1/_2$	9.7	13.2
6	11.5	17.1
$6^1/_2$	13.5	20.0
7	15.7	23.3
8	20.5	30.4
9	25.9	38.4
10	32.0	47.5
12	46.2	68.5

Crystalline TNT (flaked and grained) is used in the manufacture of cast Pentolite boosters (see Glossary), in certain slurry explosives, and in some blasting accessories.

Most forms of TNT are packed and shipped in 60 lb fibreboard cases with polyethylene liners, but special packs are used for seismic exploration applications.

Dynamites and Gelatines

Most dynamites employ nitroglycerine (NG), with some nitroglycol, as the explosive base. Gelatines also contain nitrocellulose (NC). Some manufacturers (Trojan) use nitrostarch as the base.

Nitroglycerine (NG) is the most basic of all industrial explosives, although ammonium nitrate (AN), as the oxidizing element in blasting agents, has recently gained tremendous significance.

Nitroglycerine is a yellow oily transparent liquid made by the action of nitric acid on glycerine. In this (liquid) form it is too sensitive to be handled safely; it is therefore converted into a more convenient gelatinous (plastic) solid by the addition of 8 per cent guncotton (NC) to form Blasting Gelatine (in accordance with NOBEL's patent of 1875); or by admixtures with other explosive agents and additives (to form lower strength dynamites and gelatines) in line with NOBEL's invention of 1879.

Nitroglycerine explosives require careful handling. They are rated among the most powerful of the industrial explosives. They are easily ignited, in which case they will burn slowly in small quantities but may detonate if burned in a mass.

Nitroglycerine freezes at 55 °F as a crystal. In thawing out of this phase, it is sensitive to shock. In order to reduce this hazard, all NG-based high explosives in the North American cold climate incorporate in their formulation a proportion of explosive compounds which lower the freezing point.

Nitroglycerine (and most other nitro-compounds) are poisonous if taken internally. Contact with the skin or inhalation of the vapors (except nitrostarch) will give rise to "powder headache". The hands should be washed after handling explosives.

Dynamites in North America are manufactured in the following classifications:

Straight Dynamites
Ammonia Dynamites
Straight Gelatine Dynamites
Ammonia Gelatine Dynamites
Semi-gelatines.

Practically all dynamites contain NG in their formulations. All are detonating explosives; and for proper and safe control, must be initiated by some form of blasting cap (detonator) or detonating cord (see Glossary).

Straight NG dynamites

Guhr Dynamite, as originally formulated by NOBEL in 1867, by absorbing liquid NG in kieselguhr to form a plastic mass, was the fore-runner of the straight dynamites. As such, it contained about 75 per cent NG.

Present-day straight dynamites have various proportions of active constituents substituted for much of the kieselguhr with resulting higher performance. They also use low freezing explosive compounds as replacements

TABLE 2
Compositions and Properties of Dynamites

Item	Unit	Dynamites		Gelatines		Semi-gelatines
		Straight	Ammonia (High-density)*	Straight	Ammonia	
COMPOSITIONS:						
Nitroglycerine (NG)	% by weight	20.2-56.8	12.0-22.5	20.2-49.6	22.9-35.3	see later
Sodium nitrate (SN)		59.3-22.6	57.3-15.2	60.3-38.9	54.9-33.5	
Ammonium nitrate (AN)		-	11.8-50.3	-	4.2-20.1	
Nitrocellulose (NC)		-	-	0.4- 1.2	0.3- 0.7	
Carbonaceous fuel		15.4-18.2	10.2- 8.6	8.5- 8.3	8.3- 7.9	
Sulphur		2.9- 0	6.7- 1.6	8.2- 0	7.2- 0	
Antacid		1.3- 1.2	1.2- 1.1	1.5- 1.1	0.7- 0.8	
PROPERTIES:						
Weight strength	%	20-60	20-60	20-90	30-80	63
Cartridge strength	%	20-60	15-52	30-80	35-72	30-60
Cartridge count	per 50 lb case	100-106	110	85-105	90-106	150-110
Specific gravity	-	1.4-1.3	1.3	1.7-1.3	1.6-1.3	0.9-1.3
Confined VOD	ft/sec	9000-19000	8000-12500	11000-23000	14000-20000	10500-12000
Water resistance	-	poor-good	fair	excellent	excellent	fair-v/good
Fume class	-	3	1	3-1	1	1

* Low density ammonia dynamites have a weight strength of 65%, cartridge strengths of 20-50%, cartridge counts of 174-120, specific gravities of 0.8-1.2, and two velocity ranges 6300-8100 and 8300-11000 fps. Water resistance is poor-fair, with fume class of 1.

for much of the NG. [It will therefore be seen that the term "Straight Nitro-glycerine Dynamite" is no longer straight]. Table 2 gives a range of com-positions and properties derived from DICK[1], representing average formu-lations of high explosives marketed by various manufacturers.

As will be seen from Table 2, straight dynamites achieve a high VOD and therefore give fast shattering results. They generally have fair water resistance, but poor fuming characteristics, which disqualifies them for underground applications.

Because of their high cost, their sensitivity to friction and shock, and their high inflammability properties, their industrial use and importance is declining and giving way to ammonia dynamites. Their main usage is for ditch-blasting by the propagation method (see Chapter 21).

Ammonia dynamites

Ammonia Dynamites (also known as "Extra Dynamites") have largely replaced straight dynamites in the commercial field and are in fact one of the most commonly used dynamites in presentday operations.

Although similar in composition to straight dynamites, a portion of the NG and SN is replaced by AN. This means that the ammonia dynamites generally are of lower VOD, lower density, higher shock resistance, and better fume characteristics.

They are used for rock of medium hardness and average water condi-tions in quarrying, stripping and some underground mines, in small diameter holes. They are also used for small blasting operations and agricultural purposes.

Some manufacturers produce both high- and low-density ammonia dynamites, and high- and low-VOD versions of the low density product. Low-density ammonia dynamites are generally the lowest cost cartridged explosives available. They contain rather less NG and more AN than a typical high-density ammonia dynamite. Since they have poor to fair water resistance, they appear to have less to commend their use than blasting agents (see later).

Straight gelatine dynamites

The most distinguished member of this class of high explosives is Blasting Gelatine, invented by ALFRED NOBEL in 1875 by incorporating 92 per cent of liquid NG with 8 per cent of nitrocellulose.

With a weight strength of 100 per cent (comparable with TNT) and a specific gravity of 1.3 (cartridge count, 110), it has a cartridge strength of 90 per cent.

Although it detonates at a very high velocity (25,500 ft/sec) and has excellent water resistance, it has very poor fume characteristics and cannot be used underground. Its chief industrial use (despite its high cost) is in high pressure under-water operations.

Lower strengths of straight gelatine dynamites contain NG gelatinized with nitrocellulose (guncotton) as the explosive base; they therefore have excellent water resistance and a high VOD and density.

Formerly, straight gelatines were used in extremely hard rock or in under-water and oil well shooting. They are now used for the latter purpose and for seismic blasting.

However, it is important to note that straight gelatines may not develop their fully rated VOD if the degree of confinement and the initiation energy are inadequate; or in cases where high hydrostatic pressure exists. In order to avoid this problem, a high-VOD series of straight gelatines has been developed. These are more sensitive to detonation shock, and always achieve their rated VOD, even under high water pressure.

Ammonia gelatine dynamites

An ammonia gelatine (or "gelatine extra") bears the same relationship to straight gelatines as ammonia dynamites bear to straight dynamites, in that part of the NG and SN has been replaced by AN.

Ammonia gelatines have generally a lower VOD, lower water resistance, better fume characteristics, and lower cost than comparable grades of straight gelatines (see Table 2).

Semi-gelatines

This type of "hybrid" explosive is designed to bridge the gap between high-density ammonia dynamites and ammonia (or extra) gelatines. It is a good general-purpose cartridged high explosive with considerable economic advantages, suitable for use underground. Its properties are listed in Table 2.

Further background information on the rationale of semi-gelatines is given by FORDHAM[2]:

> "The true [NG] powder explosives contain no nitrocellulose and therefore, can only be made at relatively low densities and are also susceptible to the action of water. The true gelatine explosives, on the other hand, have a continuous phase of gelled nitro-glycerine and, therefore, have a high density and are relatively unaffected by water for appreciable length of time. Semi-gelatine explosives can be made with properties which range over the extreme limits between powders and gelatines."

> "The choice of composition of a semi-gelatine depends ultimately on two requirements, namely, the strength required and the resistance to water needed for the particular application. For economic reasons the lowest nitroglycerine content which satisfies both these requirements is always chosen."

Trade names of dynamites and gelatines listed by various manufacturers are given in Table 3.

Slurry Explosives

During the last few years the extended use of slurry explosives (as distinct from slurry blasting agents — see Chapter 4) has gained in significance, and now warrants a place as a separate category in the family of high explosives. Although TNT and the dynamites develop their explosive power on detonation from a basic decomposition reaction, resulting in a transformation of their molecules, slurry explosives (SE), in common with the blasting agents, are based upon a different type of action. Their main explosive base is ammonium nitrate (in one form or another), which is an oxidizer. It depends on the complementary action of a reducing agent to support or provide the chemical (explosive) reaction. Such reducing agents are termed **sensitizers,** and generally comprise a range of carbonaceous combustibles or fuels, TNT, smokeless powder, finely divided weak metals (such as aluminium), or certain organic compounds.

It is important to note that slurry explosives and slurry blasting agents (see Chapter 4) are similar in many respects. Both are AN-slurries; both have excellent water resistance and low initiation sensitivity; both have similar ranges of density and detonation velocity; and both respond to metallization.

TABLE 3

Trade Names of Dynamites and Gelatines Listed by Various Manufacturers

(Many of these products are marketed in several different grade strengths, designated by code numbers and/or letters)

Manufacturer	Straight dynamite	Ammonia dynamite		Straight gelatine	Ammonia gelatine	Semi gelatine
		high density	low density			
Apache	"Straight dynamite"	"Standard dynamite" "Special bag"	"Special dynamite" "Special bag"	"Gelatin"	"Special gelatin"	"Amogel"
Atlas	–	"Extra dynamite"	"Ammodyte"	"Giant gelatin"	"Power primer"	"Gelodyn"
Austin	"Ditching dynamite"	"Red Diamond"	–	–	"Red Diamond"	"Apcogel"
CIL	"Ditching dynamite"	"Dynamex"	"Belite" "Blastol" "Stopeite" "Loshok"	"Giant gelatin" "Hi-Velocity gelatin" "Submagel"	"Forcite" "Powerfrac"	"Dygel" "Cilgel"
Hercules	"NG dynamite 50%"	"Extra dynamite"	"Hercol" "Hercon" "Hercomite"	"Hercules gelatin" "Herc HP gelatin" "Gas well gelatin"	"Gelatin Extra" "Power Gel"	"Gelamite"
Independent	"Straight dynamite"	"Super dynamite"	–	"Straight gelatin"	"Super gelatin" "Mighty Primer XI"	"Mighty Gel" "Genne Gel" "Unitegel"

In addition, Trojan manufactures a range of high explosives based on nitrostarch. Apart from "Super dynamite", these are difficult to classify in this table. Such products include Trojan Standard, WR, and Trojamite.

The chief and necessary distinction lies in the nature of the sensitizer used.

The employment of any such sensitizer that is inherently cap-sensitive, or promotes initiation sensitivity in the slurry mixture to the point of cap-sensitivity, will result in the mixture being classed as a slurry explosive, in the Explosive Class A category. Otherwise, it may be ranked in the less restrictive category of slurry blasting agents (SBA) (see Chapter 4).

Historically, slurry explosives were developed by COOK and FARNAM when conducting field experiments in Canada in the 1950's to produce a blasting agent with adequate water resistance.[3] They dissolved AN in just enough water to provide a supersaturated "soupy" mixture (slurry), and added TNT and other alternatives to act as a sensitizer, thereby to develop a detonable mixture which, when used with a thickening agent, possessed great water resistance.

Slurry explosives are essentially AN slurries sensitized with high explosives (such as TNT) or with certain organic compounds.

Some manufacturers use the term 'water gels' for what are generally known as slurries.

Trade names of slurry explosives marketed by different manufacturers are shown below.

Apache:	"Dynagel"
Atlas:	"Aquagel"
	"Powermax"
CIL:	"Nitrex"
	"Powermex"
DuPont:	"Tovex"
	"Drivex"
Gulf:	"Slurran"
	"Detagel"
Hercules:	"Gel-power A"
IRECO:	"IRETOL"
	"IREMITE"

High Explosives for Particular Applications

Certain high explosives have been adapted for particular applications, such as:

Coal mining (see Chapter 20)
Seismic exploration (see Chapter 14)
Overbreak control (see Chapter 14).

Details are given later in the chapters referred to.

1. Permissible (permitted) explosives

Permissible explosives are designed for use in underground coal mines to avoid methane/coal dust explosions (see Chapter 20). In order to meet the exacting demands of the U.S. Mining Enforcement and Safety Administration (MESA), they must be manufactured within especially close tolerances.

The current active list of (approved) permissible explosives published by MESA includes 70 explosives, 50 of which are of the granular type, 12 are gelatinous, and eight are permissible slurries (water gels).[4]

These are supplied by six manufacturers whose products are named below; but some lines may have since been discontinued.

Manufacturer	Granular type	Gelatinous type	Slurry type (water gels)
Atlas	Coalite	Gelcoalite	Atlas 6-B, 6-F
Austin	Red Diamond	Red-D-Gel B	—
DuPont	Duobel	Gelobel AA	EL-800B, 812,
	Monobel		816, 837
	Lump Coal CC		Tovex 300, 305
Hercules	Collier C	Hercogel A	—
	HP 226	HP 299	
	Red HA etc.		
Independent	Independent A-K	Independent Gel-A	—
Trojan	Super	Super Gel	—

The granular types are suitable for use in relatively dry coal mines. They have a VOD ranging from 5200 to 10,700 ft per second.

For wet coal mines or for rock blasting in coal mines, the gelatinous varieties are better suited. This type has a VOD ranging from 9800 to 17,000 ft per second. Cartridge counts ($1^1/_4$ x 8) range from 125 to 96 per 50 lb case.

Permissible slurry (water gel) explosives do not contain NG. They are suitable under wet conditions, and have detonation velocities ranging from 10,000 to 11,600 ft per second, with cartridge counts ($1^1/_4$ x 8) from 162 to 125 per 50 lb case.

In Canada, "permitted" explosives are manufactured by CIL as set out below, and approved by the U.K. Ministry of Energy.

Item	Cartridge count ($1^1/_4$ x 8)	VOD (ft/sec)
GELATINOUS TYPE		
C-X-L-ite	96	16,000
AMMONIA DYNAMITE TYPE		
Monobel No. 4	129	9,000
Monobel No. 7	146	—
Monobel No. 14	205	6,000
Monobel X (Eq. S)	130	6,500

2. Seismic explosives

In seismic exploration work for mineral discovery, shots are fired in the earth's crust, often under high heads of water, so that the resulting ground vibration readings can be recorded by geophones and seismograph in selected locations (see Chapter 14).

The charging of seismic holes therefore calls for special aids and sometimes for explosives of excellent water resistance. A very wide range of conditions is encountered both on land and offshore on the continental shelf.

For the severest conditions, some form of straight gelatine or slurry explosive is used, tapering off to high ammonia dynamites, semi-gelatines (and even blasting agents, see Chapter 4, in the name of economy when justified by easier hole conditions).

Common to all seismic hole charging methods is the need for continuous coupling of cartridges or cans to load them into the holes without damage, often under high heads of water (see Chapter 11). The use of explosives in seismic geophysical surveying is a specialist operation.

Trade names of seismic high explosives prepared by different manufacturers are given below.

Apache:	Seismograph High Velocity Gelatin 60 %
	Seismograph Special Gelatin 60 %
	Seismograph Amogel 60 %
	Seismograph Pattern Dynamite
Atlas:	Petrogel No. 1 HV
Austin:	Seismograph HV 60 % Gelatin (Austimon)
CIL:	Geogel
Hercules:	Vibrogel 3,5
	Gelamite-S
	Vibrocol 1,2
Trojan:	Seismic Hi-velocity Gelatin
	Seismic Gelatin

3. Overbreak control

For perimeter blasting, cushion blasting, smoothwall blasting, or pre-shearing, a special range of explosives, packaging and loading equipment has been prepared (see Chapter 14).

Most of these cartridges are designed to reduce overbreak by giving a poor degree of *coupling* (see Glossary). They are generally of the order of 5/8 to 7/8 inch in diameter.

Trade names of these special blasting supplies are given below.

Apache:	"Smoothite"
Atlas:	"Kleen-Kut"
Austin:	"Red-E-Split"
CIL:	"Xactex"
	"Shearex"
DuPont	"Tovex T-1"

Hercules: "Hercosplit"

Independent: "Split-Ex"

Trojan: "PSP" (based on nitro-starch)
"Tru-cut"

General

Descriptions of packaging methods for general and specific purposes are dealt with in Chapter 11. Methods of loading charges are also given in Chapter 11.

Purely as a guide to readers, certain prices of explosives are quoted, as an approximate average figure. These do not apply to any particular supplier; and, in any case, are subject to periodic change.

The explosives prices quoted below are (for 1971) in 20-ton car-load lots, in $1\frac{1}{4} \times 8$ inch cartridges, and delivered to states in close proximity to the factory. For more remote states, for smaller or larger cartridges, and for smaller quantities, etc., higher prices or surcharges may apply.

Explosive type	Average price per 100 lb (US $)
Straight dynamites	31.75
Ammonia dynamites	24.00
Straight gelatines	28.25
Ammonia gelatines	26.00
Semi-gelatines	25.00
Permissible (granular)	18.75
Permissible (gelatinous)	22.50
Seismic	30 to 50
Overbreak control	32 to 39

Obviously, these prices need to be escalated for 1978 conditions, but they should form a useful guide to the relative prices between each type.

References

1. R. A. DICK, "Factors in Selecting and Applying Commercial Explosives and Blasting Agents", *U.S.B.M., I.C. 8405*, 1968.
2. S. FORDHAM, *High Explosives and Propellants*, (Oxford, Pergamon, 1966).
3. B. G. FISH, "Slurry Explosives and their Application," *The Quarry Managers' Journal*, Feb. 1972.
4. J. RIBOVICH, R. W. WATSON and J. J. SEMAN, "Active List of Permissible Explosives and Blasting Devices Approved Before December 31, 1975". *Mining Enforcement and Safety Administration, I. R. 1046*, 1976.

CHAPTER 4

Blasting Agents

History and Development

Although originally patented in 1867 by two Swedish scientists (OHLSSON and NORRBEIN) as an explosive mixture, the use of ammonium nitrate, sensitized with a carbonaceous additive, did not develop commercially until 1955. Meanwhile, AN itself had been used extensively as a component of NG explosives to provide lower-strength types at lower costs; and in about 1935, a pre-packaged form of AN blasting agent was developed and marketed by DuPont[1] under the name of "Nitramon". It was sealed in metal cans to render it waterproof. A high explosive primer containing TNT was necessary for initiation. Other associated derivatives were later developed.

After the Second World War (when the need for military explosives had been greatly reduced), the United States Government encouraged the development of a non-caking grade of ammonium nitrate (FGAN) for fertilizer purposes in order to absorb the excess capacity.

Following disastrous explosions of shiploads of FGAN (unwittingly sensitized with a wax coating as an anti-caking agent) in Texas City, in Brest (France), and in the Black Sea in 1947, it became obvious that a new explosive of tremendous power was available to be developed. Various field experimenters followed up this lead in an effort to effect economies in their mining operations.[2, 3]

By 1955, R. I. AKRE, drilling superintendent at Maumee Collieries Co., Terre Haute, Indiana, and HUGH LEE had patented the use of this low-cost FGAN (mixed with carbon black) as "Akremite". Others had been experimenting with fuel oil as a sensitizer for FGAN. By 1956, Professor COOK was able to show that a 94/6 mixture of AN and fuel oil was an approximately correct composition on an oxygen-balanced basis to yield a maximum of energy from the explosive reaction.[3] A worldwide revolution in explosives usage (for nonmilitary purposes) has been developing since then.

However, AN/FO, as a dry mixture had one major disadvantage. It had no water resistance. Although its use in industry had gained tremendous momentum because of its safety and economic advantages, it could not be used satisfactorily under wet conditions. Cook and others were inspired to experiment further to overcome this problem. By 1960, they had dissolved the AN in just enough water to form a soupy mixture or "slurry". By adding TNT and aluminium powder as sensitizers, they developed a detonable mixture which, when used with a thickening agent, had great water resistance.[3] Since then, slurry blasting agents and explosives have been further developed and have become of major significance in the explosives industry.[3, 4]

General Composition

This family of blasting agents represents simply a mixture of ammonium nitrate, as a powerful but relatively insensitive oxidizing agent, and some form of combustible, usually carbonaceous.

Some of the carbonaceous "sensitizing" agents used are: carbon black, powdered coal, sawdust, sugar, molasses, and more generally, No. 2 diesel or home heating fuel oil.

The rationale of blasting agents is described in Chapter 2, giving a definition (by the National Fire Protection Association) and a classification for its safe storage and transport by the U.S. Department of Transportation.

To meet these prescriptions, all blasting agents and each of their components must be cap-insensitive, *i. e.,* incapable of being detonated when initiated by a No. 8 strength blasting cap.

In Canada, blasting agents are classified as Explosives Class 2. But even though not classified as explosives in the United States, it is convenient to regard them as explosives in the broad general sense of the term.

Advantages of a Blasting Agent

The chief advantages in its use are related to economy, efficiency, and safety. In certain applications, an overall cost saving of up to 75 per cent over conventional NG explosives has been reported. Where used under well-controlled conditions, it is reported to perform as well as or better than the dynamites, and, by virtue of its greater gas production, may even give better fragmentation. It is safer to handle and use by reason of the fact that its hazard sensitivity is low and misfires are easily and safely resolved.

One of its important virtues is that it is not classified as an explosive; but, when mixed in the correct stoichiometric proportions under preferred physical conditions with adequate primage, the mixture becomes a powerful low-cost explosive.

Disadvantages

Ammonium nitrate, being soluble, is not resistant to water—hence, there is a loss of efficiency or failure when AN/FO (see later) is used under wet conditions. (On the other hand, specially designed AN-slurry mixtures have been developed for blasting in wet holes.[2])

Because of the widely varying conditions and equipment available at different mining sites, non-standardized mixing methods may give rise to loss of control and low efficiency. As with all "do-it-yourself" systems, the results are only as good as the methods used.

Compounding of Blasting Agents

The commonest carbonaceous fuel in general use is about 6 per cent by weight of fuel oil; but an intimate degree of mechanical mixing is required in order to effect an efficient reaction; otherwise reddish-brown fumes of oxides of nitrogen and poor blasting performance result.

Unfortunately, AN particles collect moisture and tend to cake. For this reason, FGAN particles were formerly protected with wax coatings, until several ship-loads "blew up" with disastrous results. They were later coated with inert diatomaceous earth (kieselguhr).

Whereas granular particles of AN would not readily or evenly absorb oil, AN is now manufactured in porous "prilled" forms by spraying a concentrated solution through coarse spray nozzles into a stream of cool air. Prilled AN, coated with kieselguhr, readily absorbs by a sort of capillary diffusion process about 6 per cent of fuel oil. However, the kieselguhr coating is inclined to inhibit initiation sensitivity. Some suppliers overcome this problem by using surface-active agents instead of kieselguhr in the production of prilled AN.

The difficulty in securing an intimate mixture of AN and fuel oil is perhaps because they are mutually repulsive. HINO and YOKOGAWA[5] have shown that this problem can be ameliorated by the use of surface-active agents.

COXON[6, 7] has shown by practical trials that the physico-chemical bond between AN and fuel oil can be substantially improved by use of surfactants and water which produces emulsions of fuel oil in water, resulting in greater sensitivity and better explosive properties.

For dry holes, 94/6 AN/FO seems to be the best mixture yet developed. Where wet conditions prevail, a series of AN-slurry mixtures has been developed by Professor COOK[2, 3] and others. These consist generally of AN-TNT-water or AN-molasses-water. Metallized slurries have also been developed (see later).

Performance Parameters

Basically, the performance of blasting agents depends on the intimacy of the mixing process of the two main components. This, in turn, depends on the physical characteristics of the AN in respect of such features as size, shape, grading, hardness, porosity, and moisture-resistance of particles, and whether they have been treated with inert coating agents or surfactants.

Suppliers of AN blasting agents now aim to incorporate the best of the above features into their (prilled) products under competitive marketing conditions. Prills give a lower charge density and better oil absorption, thus improving initiation sensitivity and blasting performance. For effective performance, the mixture should release a maximum amount of available energy. This infers a high performance sensitivity (for both initiation and propagation), high borehole pressure, and a high VOD (both transient and steady-state).[2, 3, 4]

Other important parameters are (a) the diameter of the charge or borehole, (b) the loading density of the charge (which for AN/FO should not exceed about 1.1 gm/cc), (c) the degree of confinement of the charge, (d) the correct stoichiometric addition of sensitizer, (e) the absence of excess moisture, and (f) strength, shape, and position of primer or booster.[8]

As with all blasting agents, adequate primage with high-VOD primers is essential. Inadequate priming does not allow a blasting agent to develop its rated VOD. The reaction tends to fade, part of the charge may misfire, and noxious fumes may occur.

Types of Blasting Agents

In further describing the broad spectrum of blasting agents in this chapter, it will be convenient to make two separate classifications, as set out below.

1. Dry free-running blasting agents

These are basically prilled AN or a mixture of AN prills and fuel oil (AN/FO). They may be purchased separately or ready mixed, in 50 or 80 lb polyethylene or multi-wall bags, or in bulk.

For relatively dry holes they have a great advantage over cartridged or canned explosives in that better coupling is obtained. For quarry applications, the holes can be loaded by gravity (simple pouring) or by bulk trucks; or by pneumatic loaders for underground work (even for long "up" holes). See Chapter 11 for further details.

The most distinguished member of this family is AN/FO. Proper mixing of the AN and the oil is essential. When gravity loaded into holes, the density varies between 0.75 and 0.95 gm/cc for best results. AN/FO has poor water resistance. To overcome this problem, some quarry operators line the hole with a plastic sleeve and load the AN/FO into the sleeve (Fig. 1).

The confined VOD of AN/FO depends upon the diameter of the hole, even with adequate primage. Table 4, derived from DICK[9], demonstrates this feature.

TABLE 4
Confined Detonation Velocity and Borehole Loading Density of AN/FO

Borehole diameter (inches)	Confined VOD (ft/sec)	Loading density (lb/ft of borehole)
1.5	7000–9000	0.6– 0.7
2	8500–9900	1.1– 1.3
3	10,000–10,800	2.5– 3.0
4	11,000–11,800	4.4– 5.2
5	11,500–12,500	6.9– 8.2
6	12,000–12,800	9.9–11.7
7	12,300–13,100	13.3–15.8
8	12,500–13,300	17.6–20.8
9	12,800–13,500	20.0–26.8

Y<small>ANCIK</small>[10] has shown that at least 25 variables affect the blasting performance of AN/FO. But for a loading density of 0.85 gm/cc and under other controlled conditions, he shows that:

1. The theoretical energy released (heat of explosion) and the VOD are both at a maximum at the point of oxygen balance (i.e., at 5.7% fuel oil). The values of both factors fall off more gradually with an over- than with an under-supply of oil.

2. Initiation sensitivity maximizes at a two per cent fuel oil content but falls off rapidly above six per cent.

3. Both the initiation sensitivity and the VOD increase with the degree of confinement and borehole diameter (up to 6 and 11 inches diameter respectively).

Fig. 1. Auger-loading dry AN/FO into sleeve-lined hole from a mix truck.

4. The VOD increases and the initiation sensitivity decreases markedly when the loading density exceeds 0.9 gm/cc. This depends partly on the particle size range. Ejector type pneumatic loading machines may sometimes compress the charge to a density of 1.2 gm/cc at which point the detonation may not be sustained, even though adequately primed. To avoid this problem, ejectors should be operated at lower air pressures (see Chapter 11).

5. With finer particle sizing than —8/+20 USSS, the initiation sensitivity increases markedly. However, other physical problems then arise.

6. The length of exposure time to water is more effective in reducing VOD and initiation sensitivity than the actual water content. The blasting performance of AN/FO is not reliable in the presence of water.

7. The critical diameter of holes charged with AN/FO depends upon the degree of confinement and the loading density. With good confinement, AN/FO can be successfully fired in boreholes as small as 1 inch in diameter. As the loading density increases, so does the critical diameter. At a packed (limiting) density of 1.25, the charge may not detonate, even though the borehole diameter may be six inches or greater. With pneumatic placement in small diameter holes, high densities should be avoided.

8. The VOD decreases as the coupling ratio (or degree of coupling) falls off. This becomes more pronounced in the field of overbreak control, a procedure that capitalizes on this relationship.

9. Standard 50 gr/ft detonating cord should not be used with AN/FO in blastholes less than six inches in diameter. Such a practice promotes low order detonations laterally and tends to nullify the effect of the primers in the column.

10. High energy primers are essential for use with all blasting agents.

11. The diameter of the primer should approximate the hole diameter for maximum detonating effect in most blasting agents.

12. The use of AN/FO in small diameter underground blastholes ensures better coupling through pneumatic placement than with cartridged explosives. However, care should be taken to limit density to 1.1 gm/cc.

The classification of these mixtures as blasting agents by the Department of Transportation means that the components can be delivered separately to the blast site (without incurring excessively costly and restrictive transport and storage conditions) and mixed locally as and when required. Alternatively, they can similarly be supplied premixed (at the factory), usually in 50 or 80 lb polyethylene or multiwall paper bags.

With the availability of AN prills and AN/FO mixtures, most opencut mines and quarries have discontinued using cartridged NG explosives (except for special purposes); many underground mines are gradually following this practice, with consequent significant improvement in operating costs.

Under wet conditions and for special purposes, however, some operators prefer to retain the use of conventional explosives. Slurries are now being adapted for general underground usage where wet conditions prevail.

In very hard rock that does not readily respond to the low loading density of AN/FO charges, the use of aluminized AN/FO products may be indicated. In these formulations, either powdered or granular aluminium is added to increase the density and available energy of the charge.

2. Slurry type blasting agents

The lower costs and desirable safety aspects achieved by the development of AN-based blasting agents inspired researchers to solve their notorious disadvantages of low water resistance[3, 11, 12, 13].

This work has led with signal success to the development of *Slurry Blasting Agents* (SBA) in which the combustible fuel (as a sensitizer) is mixed with granular AN (and sometimes SN) and dispersed with just enough of an aqueous solution of AN to give a thick soup mixture. Hence the term *slurry*. They contain up to 20 per cent water.

Several types of combustible fuel have been used, e. g. molasses, sugar, sawdust, sulphur, or a heat-producing metal such as magnesium or aluminium; or even TNT (very successfully), but in this case the mixture should be referred to as a *Slurry Explosive* (SE). In Canada, all slurries are classified as explosives.

In using such an SBA (or SE), the mixture is thickened with gelling agents immediately before loading in such proportions that it can be timed to form a thick gel immediately after charging into the borehole. This has the effect of immobilizing the aqueous solution and shielding the SBA mixture from external moisture, thereby ensuring almost complete water resistance.

This virtue (of promoting water resistance) is sufficient to justify the use of slurries where wet holes are encountered. However, slurries have other distinctive advantages over AN/FO, particularly where extremely hard tenacious rock formations are to be blasted.

Slurries have a higher density than AN/FO, ranging from 1.05 to 1.6 gm/cc. This enables the charge to sink readily (even without pumping) in boreholes containing water.

The VOD of slurries is also much higher, ranging from 11,000 to 18,000 ft per second, depending upon composition, charge diameter, degree of confinement, and density.

Slurries have similar coupling advantages to AN/FO, are less dependent upon borehole diameter, may be even more insensitive (requiring adequate high-VOD primers), but yield much higher detonation pressures (by virtue of their higher density and VOD). For this reason, borehole geometry can in some cases be expanded, and drilling costs thereby significantly lowered.

Most formulations of SBA (and SE) contain from 15 to 20 per cent water. However, by using their patented organic monomethylamine nitrate sensitizer in their Tovex products, DuPont claims that less than 10 per cent water is required, thereby increasing the available energy.

Metallized slurries are indicated where extremely hard rock (such as taconite) is to be blasted.[13]

However, where rock of lower tenacity is encountered under relatively dry conditions, AN/FO (or metallized AN/FO) is in many cases a logical choice rather than slurries, because of its superior cost advantage under these conditions. Each has its preferred field of use. Sometimes they may be used in conjunction (in the same borehole) where:

(a) intermittent charging of slurry between major runs of AN/FO will increase the borehole pressure and improve fragmentation (known as "slurry boostering"), or

(b) slurry is placed at the bottom of the hole (to counter the effects of water, or to break or adequately fragment a hard "toe"), with a top load of AN/FO.

Research work on slurries is still currently pursued. Further improvements in its use in the field will undoubtedly become manifest in the near future.

Metallizing

Both AN/FO and SBA (also SE) formulations may be enhanced in terms of explosive power by 'metallizing' with aluminium powder. An addition of 6 to 15 per cent of aluminium granules or powders can be so used to advantage in AN/FO mixtures; and up to 35 per cent in slurries. When so metallized (aluminized), the density and available energy are significantly increased, the latter due to the high heat of formation of Al_2O_3 in the reaction (13,400 BTU per lb of aluminium).

Aluminium granules and powders are specially prepared for these applications by the Aluminum Company of America (ALCOA).

Packaging of blasting agents

Where bulk supplies or bulk mixing and delivery plants are not available or cannot be justified, AN prills, AN/FO, and some slurry products can be supplied in 30, 50, and 60 lb plastic bags, or plastic-lined multi-wall paper bags, as shipped from the factory.

The early forms of blasting agents, such as DuPont "Nitramon" and the later "Nitramex 2H" were packed in metal cans to ensure unlimited water resistance. The same applies to CIL "Nitrone", "Nitrox" and "Amite".

These were originally designed for bottom loading in wet quarry holes, but the coupling qualities (see Glossary) were poor unless complemented with Pelletol or Nitropel. Their chief application at the present time is in coyote blasting (see Chapter 14); and for marine offshore seismic work (see Fig. 2).

For less severe conditions in quarry holes, a number of blasting agents are available packed in fibre drums/tubes; or in heavy asphalt laminated paper shells with metal ends; or with 23G long tapered crimp on one end and fibre insert closure on the other (see Fig. 3).

Phillips markets a special blasting agent "Philblast" in a new concept of rigid plastic tubular container to promote water resistance. See also Hercules "Herculok shells".

Fig. 2. Blasting agent in sealed threaded can units for seismic blasting.

Fig. 3. Large diameter 23G tapered-crimp-ended cartridges for quarry blasting.

The CIL product "Amite II" is supplied in a special pack of an inner and outer polyethylene bag to promote water resistance. "Metamite" is packaged similarly. It contains powdered metals which increase its density and available energy. It will sink in water and it is rated for good water resistance. "Pakamex" is a packaged AN/FO product developed for secondary blasting underground. It is packed in double polyethylene bags in $2^1/_2$, 5 and 10 lb units for chute blasting and for use in scram drifts. Another CIL packaged blasting agent is "Pyromex", designed for blasting under high temperature conditions to remove accretions from furnaces, and in hot sulphide stopes underground.

Whereas dynamites and gelatines have traditionally been supplied in rigid waxed paper cartridges of various diameters and lengths, the physical nature of AN/FO and slurries and the development of plastic films have provided an opportunity to cartridge some of these (in a range of diameters) in plastic tubing, packed in a 50 or 60 lb fibreboard carton. Such plastic shells allow for an improved coupling ratio in the blasthole.

A logical development of these plastic cartridges has recently been provided in the form of a continuous cartridge of slurry of a hose-like nature, for loading directly from a coiled tube packed in a box.

For smaller operators, slurry mixtures are available in polythylene bags that may be slit before dropping into holes in order to achieve good coupling conditions. Also **Pourable Slurries** have recently been developed to allow smaller operators to use SBA without the need or expense of a mobile mix truck. These mixtures are designed to be poured easily into a hole as small as three inches in diameter.

Further details on packaging are given in Chapter 11.

Seismic Slurries

Slurry blasting agents have been developed for particular aspects of seismic prospecting, in addition to the slurry explosives mentioned in Chapter 3.

Products Available

Representative trade names of AN prills, dry blasting agents, slurry blasting agents and seismic types marketed by various manufacturers are set out in the following table.

Manu-facturer	Prills	AN/FO	Slurry Blasting Agent	Seismic Blasting Agent
Apache	FGAN	Carbamite	Carbagel	
Atlas	AN prills	Pellite	Aquaflo Aquanal Aquaram	Petron
Austin		Austinite	Slurmite	Austimon
CIL	Prilled AN	Amex II Amite II Metamite Anfomet	Hydromex Nitrex Hydroflo	Nitrone
DuPont		Aluvite	Tovex Extra Pourvex Extra	Nitramon S, WW
ESC		Temprel	Dellek Novon MS-80	
Gulf	Gulf N-IV	Gulf N-C-N	Slurran Iron Range	
Hercules	Herco-prills	Dynatex Hercomix Tritex	Flogel Gel-power O	Vibronite Impulsar
IRECO		IRENAL	IREGEL IREMEX	
Phillips	SP2 prills	Philmix	Philpower	Philblast
Trojan		Anoil	Tromax Trojel	Hydratol

Prices

Typical comparative average prices for blasting agents in carload lots are given below, based on AN prills = 100.

AN prills	100
AN/FO mixtures	150
Slurries	185— 500
Seismic types	1000—1500

References

1. E. I. DuPont de Nemours & Co., *Blasters' Handbook,* 15th. ed. (Wilmington, DE, DuPont, 1967).

2. M. A. Cook, "Modern Blasting Agents", *Science,* Vol. 132, No. 3434 (21 October 1969).

3. M. A. Cook, *The Science of Industrial Explosives* (Salt Lake City, IRECO Chemicals, 1974).

4. R. A. Dick, "Explosives and Borehole Loading", ed. I. A. Given, *SME Mining Engineering Handbook,* (New York, AIME, 1973).

5. K. Hino and M. Yokogawa, *Ammonium Nitrate-Fuel-Surfactant Explosives. Their Fundamentals and Performance,* ed. G. B. Clark, International Symposium on Mining Research, University of Missouri School of Mines & Metallurgy (London, Pergamon, 1962).

6. R. W. Coxon, "Ammonium Nitrate Explosives — Some Experimental Mixes", paper delivered to AusIMM Annual Conference, Port Pirie, South Australia, 1963.

7. R. W. Coxon, "The Use of Surface Active Agents to Sensitize ANFO Mixtures", paper delivered to AusIMM Annual Conference, Kalgoorlie, West Australia, 1964.

8. W. C. Maurer, "Detonation of Ammonium Nitrate in Small Drill Holes", *Quarterly of the Colorado School of Mines,* Vol. 58, No. 2 (April 1963).

9. R. A. Dick, "Factors in Selecting and Applying Commercial Explosives and Blasting Agents", *U.S. Bureau of Mines, I.C. 8405,* 1968.

10. J. J. Yancik, "ANFO Manual", (St. Louis, Monsanto Company, 1969).

11. B. G. Fish, "Ammonium Nitrate as a Bulk Blasting Agent," *The Quarry Managers' Journal* (June, July, August, 1961).

12. B. G. Fish, "Slurry Explosives and Their Application," *The Quarry Managers' Journal* (February, 1972).

13. R. A. Dick, "The Impact of Blasting Agents and Slurries on Explosives Technology", *U. S. Bureau of Mines, I. C. 8560,* 1972.

CHAPTER 5

Initiating Devices

General

High explosives (and blasting agents) are designed to be relatively stable for safe handling, transport, and storage. They cannot readily be initiated by the application of heat by burning (as is the case with low explosives) because, although they will burn, their explosive power cannot be adequately developed or controlled by this method.

In order to initiate high explosives (and blasting agents), a powerful localized shock or detonation is required. This is accomplished by means of a *detonator* incorporated in a cartridge of dynamite or gelatine to form a *primer*, or, in a prefabricated initiating unit, sometimes termed a *booster*. Alternatively, *detonating cord* may be used.

All detonators consist of a metal tube or shell (of copper or aluminium) which is closed at one end and into which is pressed a base charge of PETN (penta-erythritol-tetra-nitrate) or of other initiating explosive. On this is superimposed a mixture of sensitive explosives to prime the base charge. This mixture usually contains lead azide and lead styphnate with a little aluminium powder and is referred to as the priming charge.

Plain Detonators (Blasting Caps)

This type of detonator is the simpler and is used for all general purposes under relatively dry non-gassy conditions and particularly where single independent charges are fired.

The detonator is ignited by means of a safety fuse, which is inserted into the open end as far as it will go and fixed in position by "crimping" the mouth of the tube to the fuse (see Chapter 6).

The No. 6 strength of plain detonators is in general use in North America. The tube is approximately $1/4$ in diameter and $1^3/8$ in long. They are packed in lots of 100 in a small cardboard box. They cost approximately 4 cents each in case lots of 5,000 (see Fig. 4). No. 8 plain detonators are used in cases where a more powerful detonating effect is required. They are longer than but of the same diameter as the No. 6, and have twice the strength. Both types are supplied in aluminium shells.

Plain detonators store well for long periods. However, they are readily susceptible to desensitization under damp or wet conditions. In order to prevent this they should be waterproofed after crimping.

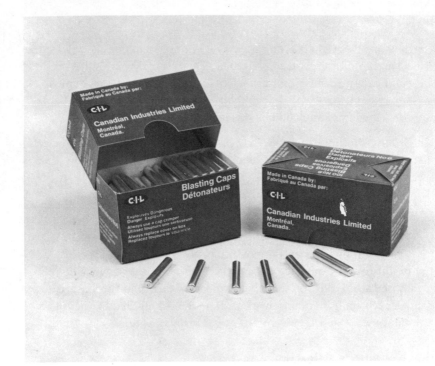

Fig. 4. No. 6 plain detonators.

Electric Detonators (Instantaneous)

These are of the same general construction as plain detonators except that an electric bridge wire is provided, and the mouth of the detonator tube is sealed with a plastic plug through which pass the plastic covered leg wires (see Fig. 5).

On the passage of a suitable electric current, the bridge wire of the fusehead "fuses", becomes incandescent and ignites the priming charge. For all practical purposes, the detonator is fired *instantaneously*, i.e., at the same time as the current is applied.

The leg wires are usually supplied in plastic covered 22 AWG tinned copper for normal applications; or with tinned iron wires for coal or salt mines; or aluminium wires for talc and white clay mining.

Cap shells are usually of copper or bronze for most applications (necessarily so for coal mining), but aluminium shells are also available. Electric detonators (EB caps) are sold with leg wires in a number of standard lengths ranging from 4 to 60 ft in steps for copper; but because of its lower conductivity, seldom more than 16 ft for iron wires. Leg wires are folded or bundled in a special way to minimize kinking when being extended during the charging process. Each bundle is plainly labelled or banded with the length of leg wire (Fig. 5).

As a safety precaution against premature explosion due to possible stray currents, the wire ends of all EB caps are completely shielded with a metal foil or plastic *shunt* which is designed to remain in place until the final act of connecting the blasting circuits is in process. Such *shunt* protection also serves to keep wire ends clean and bright until connections are actually made.

Instantaneous No. 6 electric detonators (EB caps) range in price between 25 and 110 cents each for copper leg wires ranging from 4 to 60 ft in lots of 5000.

Instantaneous EB caps are supplied in copper, bronze or aluminium shells in both No. 6 and No. 8 strengths. For seismic applications, No. 8 caps are prepared in copper or bronze shells only.

Trade names of electric detonators supplied by various manufacturers are set in Table 5. Electric delay detonators are dealt with in Chapter 9.

Detonating Cord

Another initiating device is *detonating cord* which consists of a core of initiating explosive (usually PETN) contained within a plastic sheath and wrapped in various combinations of textile, wire and plastic coverings.

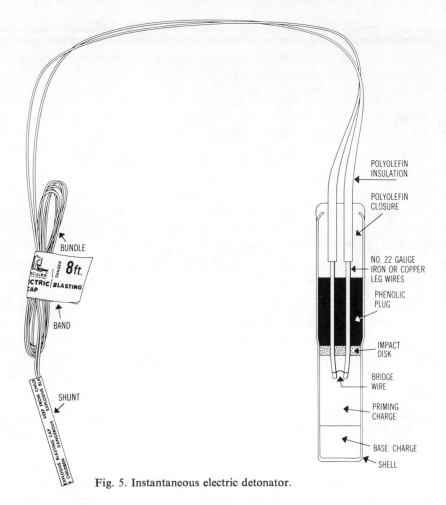

Fig. 5. Instantaneous electric detonator.

TABLE 5

Trade Names of Electric Detonators Supplied by Various Manufacturers

| Manufacturer | Classification of EB caps | |
	Instantaneous	Seismic applications
Atlas	"SF No. 6"	"Staticmaster" No. 8
CIL	No. 6	"Seismocap" No. 8
DuPont	No. 6 No. 8	"SSS Seismograph Cap" No. 8
Hercules	"Instadet" No. 6	"Vibrodet" No. 8

Although safe to handle, detonating cord (DC) can be regarded as a continuous detonator with a VOD of 21,000 ft per second, when initiated with a detonator (plain or electric) strapped or otherwise connected to the free end. To enable a network or trunkline of detonating cord to initiate each of several charges in a blast, the DC is laced through a cartridge, or threaded through a cast booster (for blasting agents and slurries) to form the primer. See Chapter 8 for further details.

Primers and Boosters

The *primer* is the key element in a charge of explosives. It is a high-density explosive unit designed to provide a high velocity and a high detonation pressure for adequate initiation of the main charge. It also provides a reliable base for locating the detonator in a firm position in the charge.

In most applications where cartridged dynamites or gelatines are used, the charge can be adequately "primed" for initiation with a primer made from one of the cartridges and a detonator.

With blasting agents, however, the initiation sensitivity is so low that a high velocity primer is required. For small diameter holes, up to $2^1/_2$ inches, a cartridge of high velocity gelatine is sufficient. But for larger diameter holes on the surface, the shape of the unit becomes an important consideration. For a given weight of the unit, the diameter is usually considered more important than its length, because it should correspond proportionately with blasthole diameter in order to project a detonation wave axially rather than laterally. Because a large diameter cartridge of gelatine would give an unnecessary length dimension, it became obvious that such priming units should be specially designed and manufactured for priming blasting agents loaded in bulk.

This provided an opportunity to prepare special factory-made cast boosters* of high velocity, and of a useful range of shapes, along with a pre-formed well to accommodate a detonator, or a small diameter hollow core through which to thread detonating cord (see Figs. 6, 7). However, only those boosters that have cap wells are recommended for use with detonators.

---

* It should be noted that, strictly speaking, a primer is an explosive unit already armed with detonator. Therefore, a booster unit, as it comes from the factory, should not be termed a "primer" until it is so armed. In the interests of safety, such loose terminology should be avoided. It is a pity that some manufacturers label their boosters as "primers".

Fig. 6. Various sizes and shapes of cast boosters.

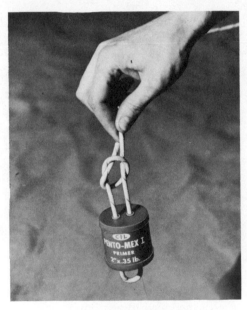

Fig. 7. Detonating cord threaded through (and looped around) a cast booster.

Many of these cast boosters are made of pentolite, a mixture of TNT and PETN, and are therefore not readily detonable by shock, heat or friction. Procore (protected core) boosters consist of TNT cast around a pentolite core.

When blasting agents and slurry explosives are cartridged in flexible plastic tubing, it is not always practicable to use cast boosters. For these purposes, a new type of booster has been developed. These may be logically termed *slim boosters* because they are prepared as a thin sheath to accommodate the detonator, which can then be entered into a plastic cartridge of AN/FO or slurry to form a primer after cutting a small opening at one end. Otherwise, they can be used in small diameter holes (less than 2 inches) pneumatically loaded with AN/FO. The term "booster" has also been used where a high velocity metallized slurry is added in a large diameter quarry hole to complement the main charge of AN/FO. This procedure is known as "boostering". In this sense, the "boostering charge" does not act as a primer, although some manufacturers unfortunately designate them as 'metallized primers'. For the sake of clarity, they are referred to in this book as "boostering charges".

For seismic blasting using a string of interlocking screwed cans of blasting agents, one such special can of the string contains a seismic booster, which has provision for arming with a seismic detonator, or detonating cord.

Trade names of various boosters marketed by different manufacturers are given in the following table.

Manufac-turer	Boostering Charges	Cast or Cartridged Boosters	Slim Boosters	Seismic Boosters
Atlas	Power primer	Power primer	Prime master	Petron
Austin		ACP primer		Austimon
CIL		A-3 primer Powertip		
DuPont		Tovex P IIDP primer	Detaprime	
ESC	Thermoprimer			
Gulf	Gulf 51B	Detagel		
Goex		Go Blast		
Hercules		Titan	Anfodet Titan EX	Vibronite S Impulsar
IRECO		IREPRIME		
Phillips		Philpower		
Trojan		NS-PD Super Prime HVT primers Prima-mite PRIMACAN	Trojan	Hydratol S

CHAPTER 6

Firing with Conventional Safety Fuse

General

This method involves the initiation of a high explosive charge by a plain detonator which has been ignited by *safety fuse*.

The basic rationale of safety fuse is the built-in safety feature provided. It is a device that burns at a closely controlled regulated rate, so that a given length represents an equivalent interval of time. In other words, a fuse of a certain length can be selected to provide adequate time for the shotfirer to retreat to a place of safety. It is of course paramount that the length of the shortest fuse used in any particular firing should provide ample time with a wide margin for safety.

Safety fuse was invented by WILLIAM BICKFORD in 1831, as a safe means of initiating black powder charges. It is still used today for this purpose. It is mainly used, when "capped" with a plain detonator, for initiating high explosives.

Safety fuse consists of a central core of specially formulated black powder formed around a centre strand of cotton, supported by and enclosed in various textile wrappings and waterproofing materials to protect it from mechanical damage and from external oil or moisture.

Some fuses incorporate a plastic sleeve in their construction. This aims to provide better resistance to water and AN/FO oil; and to mechanical damage and static electricity. Storage qualities are also enhanced.

In order that safety fuse may be readily distinguished in grey rock, black coal or in light-colored mineral headings, it is available in orange, white or black outer coverings with a heavy wax finish to provide a good crimping surface.

Safety fuse should be stored in a cool dry well-ventilated storage area, separated from contact with oils, paints, kerosene or similar solvents, which could cause deterioration by penetrating the outer covering. Fuse should be stored at temperatures between 50 and 100° F, and at low relative humidity. Older stocks should be used first.

Brands of safety fuse marketed in North America are given in the following table.

Manufacturer	Brand	Pack	Colors
CIL	Clover	3000 ft reel or 50 ft coils	black white
Coast Fuse	Plastic Sequoia	3000 ft reel	orange white black
Ensign Bickford	Sword	3000 ft reel	orange white black

Safety fuse and igniter cord cannot be used in underground coal mines.

Safety fuse is packed as one 3000 ft reel per shipping carton of 45¹/₂ lb weight; or alternatively, as 20 × 50 ft coils (1000 ft) in a 15¹/₂ lb carton (see Fig. 8).

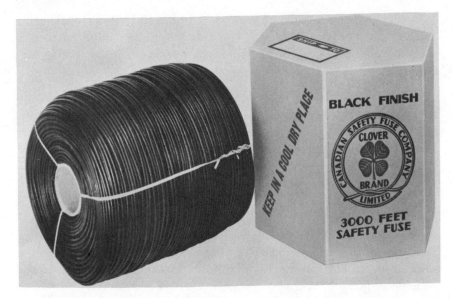

Fig. 8. Typical packages of safety fuse.

As a service to their customers, all explosives manufacturers handle orders for safety fuse and other components.

In case lots, safety fuse costs about two or three cents per ft.

Safety fuse marketed in North America is designed to burn at a standard rate of 120 seconds per yard at sea level, with an allowable variation of ten per cent either way.

With the use of safety fuse, it is of course paramount that the length of the shortest fuse used in any particular firing should provide ample time for the shotfirer to withdraw to a place of safety.

Local regulations in some cases provide that the minimum length of fuse shall be 6 ft for general blasting and 4 ft for "blockholing". Longer minimum lengths are prescribed for shaft-sinking. The breaching of such regulations is regarded by all concerned as a very serious matter.

Preparation of Capped Fuse

The crimping of safety fuse to a plain detonator (i.e., the "capping of fuse") can be performed either with a bench-mounted Detonator Crimper (Fig. 9) or with an approved type of hand crimping pliers (Fig. 10). Both incorporate a fuse-cutter and crimping jaws.

The Detonator Crimper is universally used where large numbers of capped fuses (incorrectly termed "fuse primers") are made each day in central capping stations on surface or on underground levels.

The hand crimping pliers (Fig. 10) are used for isolated *ad hoc* capping by prospectors, farmers, road gangers, and small quarry and mine operators, where a few charges are fired infrequently.

The prescribed safe series of operations for hand crimping is as follows:

1. Cut squarely across the fuse to the standard length required.

2. Remove detonator carefully from cardboard box (Fig. 4).

3. Insert freshly and squarely cut end of fuse into detonator as far as possible without using force; avoid using a twisting action.

4. Crimp mouth of detonator over fuse evenly.

5. Dip in waterproofing compound so that the whole of the detonator and about one inch of the fuse is submerged.

Fig. 9. Bench-mounted Detonator Crimper for cutting and
capping safety fuse.

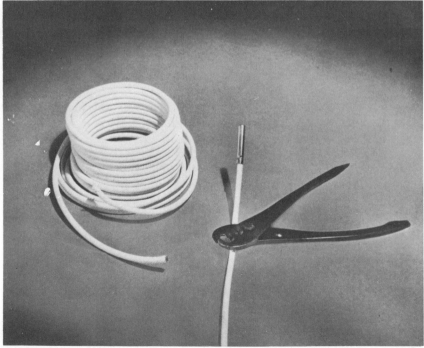

Fig. 10. Method of cutting safety fuse and crimping plain detonator to fuse with hand
crimping pliers.

6. Avoid squeezing or pressing (or applying heat to) the closed end of the detonator.

It is very important to cut the end of the fuse squarely so that the end of the powder train is in close contact with the surface of the priming charge in the detonator, as in Fig. 11(a). If the fuse end is cut on the slant, as in Fig. 11(b), the "end spit" (see later) may not ignite the priming charge and a misfire will occur.

The procedure with the detonator crimper is similar except that the detonator is inserted into the machine, the fuse is entered until it meets the charge, and the lever operating the crimping jaws is depressed. An escape chamber is usually embodied in the detonator crimper to localize safely an accidental explosion.

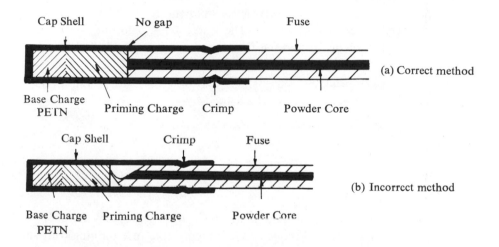

Fig. 11. Correct and incorrect methods of cutting safety fuse before crimping to a detonator.

Where igniter cord (IC) is used, the metal IC connector can also be crimped to the free end of the capped fuse (see later).

Methods of Igniting Safety Fuse

By lighting the safety fuse (with a hot flame, see later) a jet of flame known as "ignition spit" is observed by the shotfirer, giving him positive evidence that the fuse is alight and functioning normally. The burning powder then travels slowly along the length of the fuse (while the shotfirer retires to a place of safety) until it reaches the other end, which has been "crimped" into the detonator, as in Fig. 11. Here a jet of flame (referred to as "end spit") ignites the priming charge (and then the base charge) in the detonator, and the main explosive charge is thereby safely initiated.

1. **Hot wire fuse lighter:** This is a stiff wire (7, 9 or 12 inches long) coated with an incendiary composition that burns slowly with an intensely hot fire zone. This flare is held against the freshly cut end of fuse to be lighted. This type of lighter must not be used as a timing device.

2. **Pull wire fuse lighter:** The freshly cut fuse end is inserted in the open end of the lighter until it touches the end of the wire, which is then deliberately pulled, setting off a mechanical lighting arrangement.

3. **Lead spitter:** A thin lead tube filled with black powder, coiled in a 25ft length on a spool, is lighted at the free end and held against the freshly cut fuse end. The spitter gives an intensely hot flame which progressively burns the tube away. The lead tube is paid out as required and finally cut off behind the flame as the lighting operation is completed.

4. **Match, match-head, cigarette lighter:** This practice is not recommended where more than one fuse is to be lighted.

5. **Igniter cord (IC):** see below.

Igniter Cord System

Igniter cord (IC) is an automatic device for lighting any number of safety fuses in the desired sequence. It is a cord-like fuse burning with an intense flame progressively along its length at a fairly uniform rate. At certain points along its length it is connected to the safety fuses for the individual holes by a special metal connector, loaded with an incendiary composition. As the flame reaches each connector, it automatically lights each safety fuse in turn (Fig. 12).

Joints in the IC trunk line may be made by doubling the ends and tying a square knot.

Since IC gives a certain amount of time delay and may be operated from a remote point, fuses may be of shorter length; but never shorter than three feet.

The metal connectors are usually pre-crimped to the capped fuses; the IC trunk line is conveniently attached to the connectors by locating it in a special slot and closing it with moderate thumb pressure (Fig. 12).

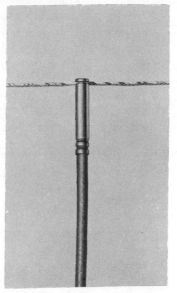

Fig. 12. Igniter cord located by thumb pressure in slotted end of metal connector which has previously been crimped to free end of capped fuse. When the flame of the burning IC reaches the connector, the fuse will be lit automatically.

Current trends show that the purchase of **Fuse Assemblies** from manufacturers will increase in future. They are factory-assembled capped fuses of various standard lengths equipped with metal IC connectors. These units are safer to handle and more reliable than field-assembled units. As this practice develops, the use of central capping stations (see above) will no doubt correspondingly wane.

Igniter cord is made in three color types with different burning rates, for particular applications:

(a) **Fast** type, burning nominally at the rate of 3—5 seconds per ft. It is colored *black*. The cord burns rapidly enough to be clear of holes already exploding, such as in narrow longwall faces, in other than coal mines.

(b) **Medium speed** type, burning at 5—10 seconds per ft. It is colored *green*. It burns rapidly enough to ignite all the fuses before the first shot explodes; and yet gives enough lead time between holes to fire in sequence (see Chapter 9). The distance between fuse connections should be two inches for every foot of fuse in a hole. Used typically in stope blasting.

(c) **Slow speed** type, burning at 15—21 seconds per ft. This type is colored *red*. It is designed for heading blasts in which many holes are to be fired in long sequences in a restricted area. Fuse connections should be spaced at a distance of one inch per ft of fuse in a hole. Sufficient time is thereby allowed to have each fuse lit before the first explodes; and yet with adequate lead time between sequential holes.

With any type of igniter cord, there is no need to trim fuses (see Chapter 9) because the firing sequence is adequately provided for in the IC circuit. On the other hand, the cord needs to be ignited at one point only, thus providing an excellent safety feature (Fig. 13).

It is important that all other fuses in the round are alight and burning before the first hole detonates. This means that the length of igniter cord

Fig. 13. Lighting one end of an IC circuit which progressively lights all the safety fuses in turn as the IC flame reaches the particular connector.

between the first and the last fuses should represent a time interval shorter than that of the first fuse between ignition and detonation. This particular length of igniter cord is known as the *limiting cord distance*. It obviously depends upon the speed of the type of igniter cord used.

All three types are marked clearly at 12-inch intervals to enable distance gauging to be readily made. Each type is supplied on 100 ft plastic reels with 50 spools (5000 ft) to a box; or alternatively, in 33 ft pull-out spool packs (Fig. 14). The cord has a finished diameter of about $1/16$-inch, and is fabricated with a core of thermite powder within wirebound textile wrappings. It is especially designed as a safe reliable method of firing a large number of holes sequentially with safety fuse in underground blasting applications other than in coal mines (Fig. 13).

Fig. 14. Igniter cord packed (a) in 33 ft pull-out spools, and (b) on 100 ft plastic reels.

Connectors are packed in boxes of 100, with 50 boxes (5000 connectors) per carton (Fig. 15).

Igniter cord is rather perishable and should not be stored for long periods, especially in a humid atmosphere. *It should never be used as a substitute for safety fuse.*

Igniter cords supplied by various manufacturers are set out below.

Manufacturer	Trade Name	Fast (3—5 sec/ft)	Medium (8—10 sec/ft)	Slow (15—21 sec/ft)
CIL	Thermalite	Black	Green	Red
Coast Fuse	Spittercord		Green (A)	Red (B)
Ensign Bickford	Ignitacord	Black	Green	Red

Average prices of fuse lighting supplies are given below.

Igniter cords (all types):	3 cents per ft.
Hot wire fuse lighters:	4 cents each
Lead spitters:	10 cents each
Pull wire lighters:	7 cents each
IC connectors:	4 cents each

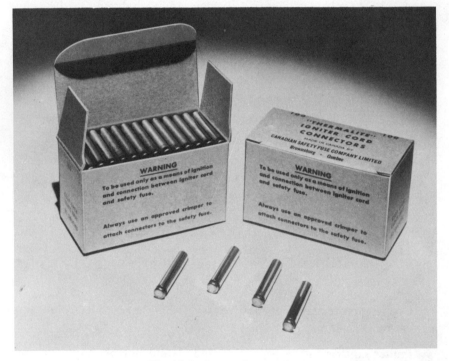

Fig. 15. Igniter cord connectors.

Preparation of a Primer Assembly

For large diameter holes on surface, and particularly where blasting agents are used, prefabricated cast boosters are employed (see Chapter 5).

These are designed and manufactured with either a cap well (to accommodate a detonator) or an axial tunnel (through which detonating cord may be threaded) or both. There is therefore no problem in activating these boosters with the necessary detonating device.

However, for the small diameter holes employed in underground blasting (and for small scale surface blasts), it is customary to form a primer from a stick of high explosive (dynamite or gelatine) of the same grade as used for charging the round; or where AN/FO is used, from the standard grade utilized on the mine (but preferably, a gelatine).

Such a primer assembly would consist of a cartridge of high explosive into which a detonator has been placed. It can be an electric detonator or a capped fuse assembly (incorrectly referred to sometimes as a "fuse primer", even though it does not become a primer until it is incorporated in a plug of explosive or a booster).

The following describes two alternative methods of assembling a primer with a cartridge of dynamite or gelatine and a capped fuse. Electric primers are dealt with in Chapter 7.

First method

1. Take a whole cartridge of high explosive; do not break it in halves; do not remove paper wrapping.

2. Form an opening about 2 in deep axially in the end of the cartridge with a wooden skewer or with a brass or copper "pricker".

3. Insert the detonator end of the capped fuse into the hole so formed until the detonator is completely buried in the explosive cartridge.

A variation of this method is first to open the wrapping on the end of the cartridge and (after the detonator has been placed) to restore the wrapping and, if desired, to tie it round the fuse with twine.

Second method

1. Take a full cartridge of explosive; do not remove wrapping.

2. With a wooden or copper "pricker" form a diagonal opening about 2 in deep in the side of and near one end of the cartridge.

3. Insert the detonator end of the capped fuse into the hole so that the detonator is completely buried.

4. Fasten the capped fuse to the cartridge with string. (For this purpose hemp twine is supplied with the capped fuse, hitched around the fuse about $1/2$ in from the crimp).

The advantage of the second method is that the fuse is not disturbed or deformed by the tamping pole when charging the hole, but the detonator cannot be placed in a truly axial position.

The main essentials of a primer are that:

1. The detonator should not become dislodged from the cartridge.

2. The detonator should be in the safest position in the cartridge.

3. The primer should be capable of being loaded safely and easily into its desired position in the hole without damage to the fuse.

4. The detonator should be inserted deeply into the cartridge (i.e., buried) and should lie as far as possible with its long axis along the centre line of the cartridge.

5. The closed end of the detonator should point towards the larger bulk of the cartridge. In other words, the mouth of the detonator should be next to the nearer end of the cartridge.

Central Capping Stations

Central capping stations are set up on medium and large mines and quarries where a large volume of capped fuse is used per day.

By centralizing these activities, the benefits of specialization are achieved. Skilled operators, and well-designed measuring, cutting, crimping, waterproofing, and coiling equipment are thus available under conditions of efficiency and safety. Routine tests can also be made of the output.

A special fireproof room, well-ventilated and secure, is constructed specially for the purpose. It embraces a receiving annexe for cases of reeled fuse, detonators, and waterproofing compounds. It also includes a storage annexe for the storage of different categories of standard lengths of capped fuse ready for issue.

The main section contains a working bench. Over the feed end is a bracket to hold a 3000 ft reel of fuse as supplied by the manufacturer, an automatic machine for measuring and cutting the desired standard length of fuse, a Detonator Crimper on which the fuse is crimped to the detonator, and a waterproofing dip tank. Automatic fuse-cutting machines at present in use have a capacity of 900 fuses per hour.

The chief advantages of a central capping station are:

1. Production of more uniformly accurate lengths of capped fuse.

2. Proper cutting of fuses.

3. More effective crimping.

4. Better waterproofing.

5. The opportunity to test.

6. Fewer misfires.

7. More efficient blasting.

8. Better supervision.

9. Improved conditions of safety.

10. IC connectors may be crimped to the free end of the capped fuse in a more efficient manner than by hand at the face.

CHAPTER 7

Electric Shotfiring

General

Interest was first aroused in electric blasting procedures over a century ago when SIEGFRIED MARKUS[1] built an electric blasting machine for "spark-detonators". In 1876, H. JULIUS SMITH invented the first electric detonator as we know it today. Electric shotfiring was introduced into the United States about twenty years later.[2]

Blasting by electrical means has great technical and economic advantages. For many reasons, firing by safety fuse and plain detonators was restricted in the past to a limited number of shots and to other than underground coal mines. With electric firing, an almost unlimited array of charges can be fired, all with a single act (of pressing a button or actuating a switch), in a safe place (remote from the blast site) and in a deliberate manner. There are other advantages, mainly in the promotion of safety.

Electric blasting methods are in general use in North America with either instantaneous or delay detonators, employing either exploders (blasting machines) or power mains as the source of the electrical impulse.

Exploder firing is relatively simple, but the number of holes per blast is limited by the capacity of the exploder.

Power mains of from 110 to 440 volts pressure are used for major blasting operations; by this method, a much greater number of holes per blast is possible. With AC mains, larger minimum currents are required than for DC circuits.

In all electrical methods, care should be taken to minimize the circuit resistance due to loose and dirty connections, and current leakage due to poor insulation.

Preparation of Electric Primers

Each electric detonator is supplied intact with plastic covered leg wires in standard lengths of 48, 72, 84, 96, 120, or 144 in and longer. The primer is prepared in the following way:

1. Take a full-sized cartridge of high explosive.

2. With a wooden or nonferrous pricker, make an axial hole in the end of the cartridge.

3. Insert the detonator in this hole until it is completely buried.

4. Secure the detonator to the cartridge by half-hitching the leg wires twice around the cartridge (Fig. 16).

5. Keep ends of the leg wires shunted and clear of all possible electric circuits.

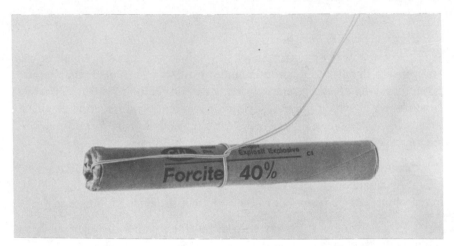

Fig. 16. Electric primer (The EB cap is located axially in one end and the leg wires are looped (half-hitched) once (as is shown) or twice around the cartridge to avoid dislodging the cap during the charging procedure).

Circuit Calculations

In any electrical blasting circuit[3], there are three basic elements:

(1) The source of electrical energy,

(2) the electric blasting caps, and

(3) the blasting circuit connecting the energy source with the EB caps.

It is customary to consider this blasting circuit in three distinct parts as set out below.

(a) The *leg wires* that are integral with the detonators. The free ends of each pair of EB cap wires are kept shunted (short-circuited) until the circuit is finally ready to be connected, immediately prior to blasting.

(b) The *connecting wires* that are sometimes necessary to extend the circuit between the leg wires and the permanent firing line. These wires are usually of 20 gauge solid soft-drawn copper, or of 18 gauge three-quarter-hard electrical-grade aluminium, either single or duplex, with vinyl insulation, packaged on metal spools. Resistances for both types average 10.3 ohms per 1000 ft (at 68° F). These connecting wires are usually damaged or destroyed at each firing, and should be long enough to prevent damage to the permanent firing line (leading wires).

(c) The *firing line*. This is not damaged when blasting from power mains is practised, and therefore a heavy gauge low resistance firing line is installed as a durable installation. Except where long distances are involved, 14 gauge copper (or 12 gauge aluminium) wires are generally used, giving a resistance of about 2.6 ohms per 1000 ft.

For certain operations, e.g., in smaller quarries, a compromise is sometimes adopted, by using an 18 gauge 2-conductor flexible stranded copper *shotfiring cord* to replace both the connecting wires and firing line mentioned above. In this case, the resistance is about 6.4 ohms per 1000 ft.

The average resistances of electric detonators used in North America can be inferred from Table 6.

In routine blasting operations, the capability of the whole circuit is usually well-established, and current practices can be successfully maintained. However, for special or trial blasts, initial calculations need to be made in order to determine the electrical resistance of the circuit and the current characteristics of the power line (or the capacity of the exploder), so that sufficient current will flow to ensure a satisfactory blast.

It will therefore be necessary to calculate the total resistance of the circuit, including that of EB caps, connecting wire and the firing line. It will also depend upon the type of circuit adopted—whether straight series, straight parallel, or parallel-series.

For small rounds, straight series or parallel circuits are used. But where a large number of detonators is involved in the one firing, a parallel-series circuit is employed.

Blasting machines (exploders) provide DC current and are best adapted for series circuits. Experience has shown that the minimum current in a straight series circuit should be 1.5 amperes DC or 2.0 amperes AC; and for series-in-parallel (parallel-series) the minimum current (DC or AC) should be 2.0 amperes in each series. For straight parallel circuits, the firing current (DC or AC) should be not less than 1.0 ampere, nor more than 10.0 amperes if arcing within the detonator is to be avoided.[3]

TABLE 6
Nominal Resistance of No. 8 Electric Detonators
(in ohms, with 23 AWG copper or iron leg wires)

Length of leg wires (ft)	Instantaneous detonators		Delay detonators	
	copper	iron	copper	iron
4	1.41	2.25	1.86	2.70
6	1.49	2.75	1.94	3.20
8	1.58	3.25	2.03	3.70
10	1.66	3.75	2.11	4.20
12	1.74	4.25	2.19	4.70
14	1.82	4.75	2.27	5.20
16	1.90	5.25	2.35	5.70
20	2.06	6.25	2.51	6.70
24	2.22	7.25	2.68	7.70

Note:

The above data are listed by Hercules[4]. Detonators of other manufacturers have different resistance values, depending mainly upon that of the bridge wire and the gauge and type of leg wires. Manufacturers use bridge wires of different resistance; therefore, it is necessary to avoid mixing detonators of different manufacture in the same blasting circuit.

Details of Blasting Circuits

1. *Series circuit:* The circuit is as shown:

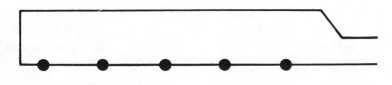

In series firing, the complete circuit can be accurately tested for continuity. The firing current should be at least 1.5 amperes in order to ensure that each detonator functions properly in a DC circuit.

As an example, let us assume that we have 50 electric delay detonators with 8 ft leg wires (assuming each of 1.6 ohms resistance) and that we are using 100 yd of 20 AWG connecting wire and 100 yd of 14 gauge firing line.

To determine the voltage required, we calculate:

Resistance of 50 electric detonators	80.0 ohms
Resistance of connecting wire $(2 \times 300 \times 10.3/1000)$	6.2 ohms
Resistance of firing line $(2 \times 300 \times 2.6/1000)$	1.5 ohms
Total resistance	87.7 ohms

Voltage necessary: $1.5 \times 87.7 = 132$ volts approx.

Series circuits are not commonly employed with mains current. They call for a low current at high voltages.

When more than 40 detonators are to be wired into a blasting circuit, it is preferable to use a parallel-series circuit. It is not advisable to fire more than 50 detonators in a straight series circuit.

2. *Parallel circuit:* In this circuit, each detonator provides an alternative path for the current.

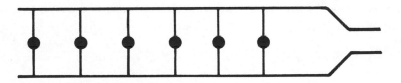

The complete circuit cannot therefore be tested for continuity, but individual detonators can be tested with an approved ohmmeter prior to charging.

For parallel firing, a minimum current of 1.0 ampere (AC or DC) is usually allowed per detonator.

Then:

Resistance of 50 detonators	0.03 ohms
Resistance of connecting wire	6.20 ohms

Resistance of firing line	1.50 ohms
Total resistance	7.73 ohms

Current: $1.0 \times 50 = 50$ amperes

Required voltage: $7.73 \times 50 = 387$ volts.

3. *Parallel-series circuit:* This is arranged as a number of series circuits connected in parallel. Each series circuit will have approximately the same resistance since it should contain the same number of detonators. No more than 40 detonators should be used in each series circuit.

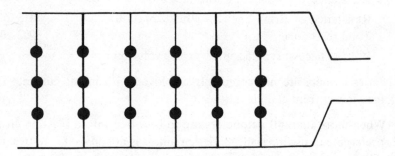

The total resistance of the circuit is given by the resistance of a series divided by the number of series in parallel plus the resistance of cable connections. If the 50 detonators are arranged in 10 rows of 5 detonators, then the current required $= 1.5 \times 10 = 15$ amperes.

Total resistance: $\dfrac{1.6 \times 5}{10} + 6.2 + 1.5 = 8.5$ ohms

Voltage required: $15 \times 8.5 = 128$ volts.

The number of series that can be used is limited by the current available, and the number of detonators in each series by the voltage.

This type of circuit is applicable to cases where 40 or more EB caps are to be fired, or where the series resistance would be unduly high, as, for example, in long-hole pattern stoping.

Shotfiring with Exploders

All exploders should be equipped with removable firing keys which are to be kept in the possession of the shotfirer.

For single shots, the leg wires of the detonators are bared and connected to the ends of the shotfiring cable. In retreat, the other ends of the shotfiring cable are connected to the terminals of the exploder which is set up at a safe distance. When all personnel have been cleared from the area, the shotfirer inserts the key and operates the exploder.

The exploder is also used to fire a multiplicity of shots, not exceeding its rated capacity.

Before firing, the electrical circuit can be proved by testing. At first, the permanent firing line is tested, (a) for continuity, and (b) for insulation failure.

When testing for continuity, the two wires at one end should first be twisted together. A blasting galvanometer connected to the wires at the other end will indicate the passage of a current and thus prove continuity; or, if an ohmmeter is used, the continuity can be tested and the circuit resistance determined directly from the ohmmeter reading.

For the insulation test, the ends of the firing line are left unconnected. The ohmmeter connected to the two wires at the other end should show that no current is flowing and that the resistance is infinite.

Next and in turn, the connecting wires and the detonator circuit are connected to the firing line and the complete circuit is tested from the firing position—not at the face. An approved ohmmeter must be used—one that employs a current sufficiently minute that risk of premature detonation during the test is avoided (see later).

For important blasting operations, it is generally considered advisable to test the circuit before every individual blast.

Shotfiring from Power Mains

Electrical energy from power mains is now commonly used for large scale blasts in underground metalliferous mines (for stoping and pillar recovery), in many large quarries and open cut mines, and for tunnelling and civil engineering construction operations.

When power mains are used, a much larger number of shots can be fired, under closely controlled conditions. Most systems utilize special permanently installed blasting circuits, incorporating locked switch boxes, firing boxes, and short-circuiting boxes to give maximum protection against premature firing (see Fig. 17). A more recent development is the provision of a master installation to serve a number of different unit stope circuits, incorporating built-in switching and circuit-testing arrangements. See Chapter 9.

Special precautions are required to prevent stray currents or accidental premature closing of the blasting switch by unauthorized persons. Switchgear is usually locked in cubicles or cabinets.

Care should be taken to ensure that the blasting switch is closed only for a moment. Prolonged closure may result in a "flashover" or arc from the brass foils of the fusehead to the cap shell, due to an excessive amount of current (exceeding about 10 amperes), thereby causing a pinhole or misfire, and perhaps prematurely igniting rather than detonating the charge.

In order to prevent or minimize this "arcing", some detonator manufacturers provide an inbuilt shield around the bridge wire of the EB cap. Others seek a limit to the blasting current by using an electronic device. Typical of these is the Tunnel Blasting Switch (Atlas).

In mines, the electric current for firing electric detonators is derived from power mains, and arrangements such as illustrated have been installed to ensure safety in use and freedom from premature blasts. In tunnelling operations also, power main firing is widely used. It is essential, with this method of firing, to incorporate as many safety features as possible to guard against lightning, stray currents, induced currents, and accidental or incorrect operation of firing switches.

Position of firing lines

The cables for blasting should be used for this purpose only, and should be carried on insulators on the side remote from pipes, other cables, ventilation ducts, and other possible conductors of stray currents.

Firing units

To include as many safety features as possible in power main firing, three units including a fuse box with main switch, a firing box, and a short-circuiting box should be used, or an arrangement giving equivalent protection against premature firing or faulty operation (see Fig. 17).

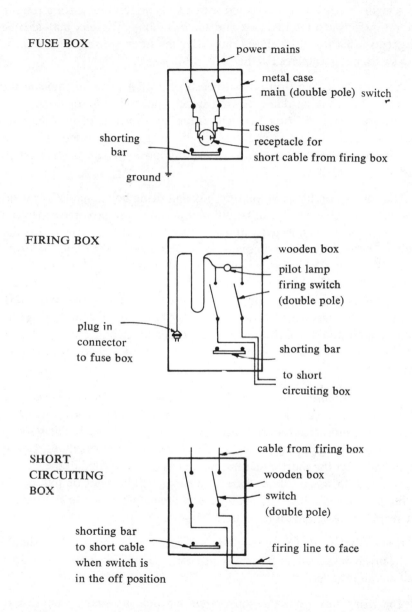

FUSE BOX

power mains

metal case
main (double pole) switch

fuses

shorting
bar

receptacle for
short cable from firing box

ground

FIRING BOX

wooden box
pilot lamp
firing switch
(double pole)

plug in
connector
to fuse box

shorting bar

to short
circuiting box

**SHORT
CIRCUITING
BOX**

cable from firing box

wooden box
switch
(double pole)

shorting bar
to short cable
when switch is
in the off position

firing line to face

Fig. 17. Power main firing units.

1. *Fuse box:* This unit consists of a grounded metal box containing fuses and a main switch, and is also fitted with a suitable receptacle for a plug-in connection between the fuse box and the firing box. The box may also be designed so that it can be closed only after the fuses are removed and the main switch has been placed in the "off" position.

2. *Firing box:* The firing box consists of a pilot light (to indicate the presence of a firing current), a firing switch, and a means for connecting between the fuse and firing box. The firing box which is of wooden construction is made in such a manner that the door can be closed only after the firing handle has been returned to the "off" position with the firing lines shorted.

The connection between the fuse box and firing box is made by means of a length of cable which can be plugged into the fuse box, the other end being permanently connected within the firing box. It is desirable that the fuse box should be situated on one side of the tunnel or drift and the firing box on the other. This arrangement ensures that there is an adequate electrical gap (lightning break) between these two units so that an electric current cannot pass from the fuse box to the firing box unless the connecting cable has been plugged in. A further advantage of the fuse box and firing box being on opposite sides of the tunnel or drift is that the connecting cable acts as an effective bar to traffic until it has been returned to its position within the firing box.

3. *Short-circuiting box:* The short-circuiting box situated between firing box and the face is an important unit which ensures that the blasting lines are disconnected from the cable and effectively shorted. The door of the short-circuiting box should be arranged in such a manner that it can be closed only after the short-circuiting switch has been returned to the "shorted" position. In this position, "stray" currents which might be induced in the blasting cables and which might cause a premature blast are shorted before reaching the detonator circuits.

It is important that the firing box and the short-circuiting box should be effectively insulated from the ground; consequently, ground connections must not be installed.

The firing line from the short-circuiting box is usually extended to about 30—40 ft from the face; from this point the connections to the face network are made with expendable, lighter gauge insulated "connecting wire". As a safeguard against stray currents, it is important that the face ends of the firing line should be twisted together to short them effectively until they are actually connected. The firing line should be insulated from the ground.

Firing procedures

After connecting the blast in parallel, series, or parallel-series circuit, personnel should retire to the firing point, leaving the foreman or shift boss, who has the only key to the firing units, to make the connection to the firing line. On his way back to the firing box, he should unlock the short-circuiting box, test the circuit through to the face, and finally close the switch. He should then retreat to the firing box, unlock it, and take out the connecting cable. After the fuse box nearby is unlocked, this cable should be plugged into the fuse box and the main switch of the fuse box switched on. This supplies current to the pilot lamp in the firing box, thus indicating that the current is available for firing.

To fire the blast, the warning siren should first be sounded; the switch in the firing box should be closed for a moment, then **immediately** withdrawn and returned to the shorted position. The switch in the fuse box is then placed in the "off" position and the short circuiting cable returned to the firing box and both boxes locked. When returning to the face, the shift boss places the switch in the short-circuiting box in the shorted position, and the box is again locked. The end of the firing line nearer the face should then be shorted by twisting the ends together and keeping it insulated from the ground. Before the next blast, the firing line should be tested to ensure that is has not suffered damage during the blast.

As the tunnel advances, the short-circuiting box, firing box, and fuse box are moved forward at intervals. In general, the short-circuiting box should be located some 200 ft back from the face, while the blasting box and fuse box should be at such a distance from the face that no danger exists from rocks projected back along the tunnel, or from air concussion.

Wiring up the blast

In tunnelling, and especially where electric storms are prevalent, special care needs to be taken to avoid premature blasts due to lightning strikes, induced currents and static electricity. It should be emphasized however, that no protection is effective against a close lightning strike, and consequently, charging operations should be suspended on the approach of an electrical storm.

Fig. 18. Generator-type blasting machines.

Fig. 19. Method of operating a No. 10
Twist-type blasting machine.

Detonators are normally supplied with the free ends of the leg wires shunted; this protection should be left in place until just prior to connecting them in the circuit.

Naturally, the ends of all wires to be connected should be thoroughly scraped and bared, with enamel insulation removed; so that they can be electrically connected with an effective twist splice.

Frequently, the parallel method of firing is employed, and under such conditions it is strongly recommended that leg wires of the same color be connected to a common "buswire" at the face. This bus-wire is later connected to the regular connecting wire of the circuit. The advantage of bus-wires is that the face circuit can readily be checked by the shift boss; otherwise, it is possible for both wires of an individual detonator to be connected to the same connecting wire; in this event, a misfire will certainly result.

Electric Shotfiring Accessories

1. **Exploders:** These are also referred to as Blasting Machines. There are two main types:

(a) The *Generator type,* operated by a rackbar or a twist spindle, to rotate the armature of a small DC generator (see Figs. 18, 19). As the generator develops full speed, a firing switch is automatically closed, thereby applying the generator current to the firing circuit to ignite the detonators; and

(b) the *Condenser discharge (CD) type,* used mainly for larger blasts. In principle, a charging button is pressed to build up a high voltage charge on a bank of condensers from a dry cell battery (Fig. 20). When fully charged, the firing button is depressed to discharge the condensers, thereby energizing the firing circuit. With later models (VME 225 A/ 450 A and BLM 100) the condensers are automatically charged.

Both types of blasting machine are designed to produce a high voltage low DC current.

Blasting machines now currently available are shown in Table 7. The ratings shown refer to machines "as new" or in good condition, and within a specific circuit resistance.

Nitro Nobel AB produces the range of blasting machines shown in Table 7 as the CI types. These are condenser discharge machines in which the condenser charge is built up with a hand-cranked generator, rather than with a battery system. Special test instruments are also available.

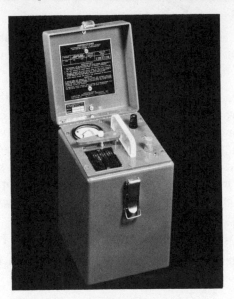

Fig. 20. A condenser-discharge
type exploder.

2. **Blasting machine testers:** It is important that exploders should be properly maintained and regularly tested as to their firing capacity. Even if well-maintained, they may develop faults in time that reduce their capacity. Blasting machines should be kept, right side up, in a cool dry place. Temperatures over 100 °F will damage the armatures of generator-type exploders.

The effectiveness of generator-type exploders (which are usually rated by the number of EB caps they are capable of firing in series, when in good condition) is commonly tested with a specially designed **Rheostat,** in conjunction with up to four caps (or preferably, electric squibs for greater convenience). The rheostat is provided with numbered binding posts corresponding to related inbuilt resistances and labelled with the number of standard caps that the exploder under test has the rated capacity to fire (Fig. 21).

A useful test circuit is shown in Figure 22 in which four caps in a parallel-series connection is wired in series with the exploder under test, and connected with the appropriate binding post on the rheostat corresponding with the rated capacity of the exploder. If all four caps explode when the exploder is operated, then the machine can be regarded as in good working order. Otherwise, it needs repair, following which it should again be tested. In the interests of safety, it is good practice to scrap blasting machines that do not develop their fully rated capacities after careful repair procedures.

TABLE 7
Blasting Machines

Name of Unit	Firing Voltage	Maximum Capacity (in caps)		
		Straight series	Straight parallel	Parallel series
GENERATOR TYPE:				
No. 10 Twist	100	10	–	–
No. 10 Twist (permissible)	100	10	–	–
No. 30	200	30	–	–
No. 50	300	50	–	3/ 40
CONDENSER-DISCHARGE TYPE:				
Handi-blaster 10	–	10	–	–
Femo (permissible)	–	10	–	–
Femco 20-shot (permissible)	400	20	–	–
Mini-Blaster	–	20	–	–
Titan	450	50	60	30/ 40
VME 225A	225	50	10	6/ 40
VME 450A	450	50	50	40/ 30
BLM 100	250	50	30	5/ 40
CIL 500	–	500	–	10/ 50
CIL 10-shot	–	10	–	–
Atlas 800C	–	50	30	20/ 40
TBS	–	50	120	30/ 40
CD 600	–	100	30	12/ 84
CI 50 2VA	340	50	–	–
CI 15VA	600	150	–	6/ 80
CI 100VA	1100	300	–	25/120
CI 275VA	2800	700	–	30/300
CI 700VA	2500	700	–	80/300
Trojan Warrior 10C	80	10	–	–
20C	–	30	–	–
40C	–	50	10	6/ 40
1000C	–	50	30	20/ 50

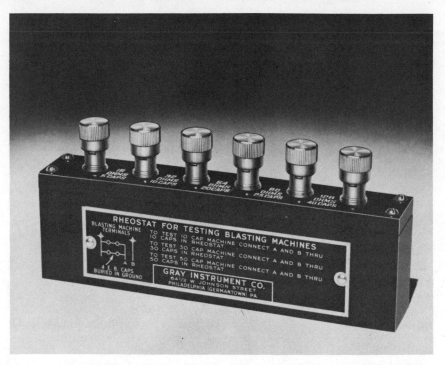

Fig. 21. Rheostat for testing blasting machines.

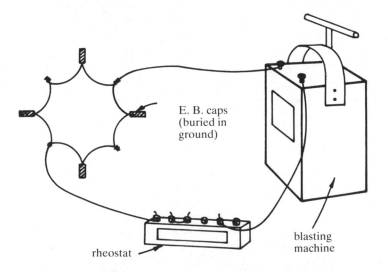

Fig. 22. Circuit for testing a 50-shot blasting machine with rheostat.

The full power of a Condenser Discharge (CD) blasting machine can be tested in the laboratory with an oscilloscope, or in the field with a voltmeter or Blaster's Multimeter[3], provided the condensers are in satisfactory condition. The CIL Performance Tester is also designed to test the condition of condensers in CD blasting machines by checking the "full power" reading of the unit under test.[6] If full power is not achieved, then the condensers are not accumulating or holding their fully rated charge; and therefore the machine must be repaired before further (checking and) use.

3. **Circuit testers:** Electric blasting procedures can be successfully executed only if proper care is exercised in planning, loading and connecting the circuit. In important blasts, where the cost of much preliminary work and explosive materials is at stake, circuits should be tested prior to every firing.

A list of possible circuit weaknesses is given below, together with a description of the instruments used to locate or test for corresponding faults.[5]

(1) *Discontinuities* or shorts in (a) the cap circuits, and in (b) the total (or any part of the) blasting circuit. A *blasting galvanometer* is used (see below).

(2) *Current leakage* can occur when damage to leg wire or connecting wire insulation allows the bare wire to make contact with the rock (ground), especially under damp conditions. This condition can be checked with a *blasting multimeter*.

(3) *Stray electricity* (current leakage from extraneous sources). Before loading the holes, the shotfirer can test for stray electricity with the probe wires of a *blasting multimeter*, see below.

A **Blasting Galvanometer** (see Fig. 23) is an Ohmmeter furnished with a silver chloride or other special cell in series with high resistance coils, designed to reduce the current to such a low value that an EB cap will not fire while testing for the conductivity of a cap circuit in series (or in any series of a parallel-series circuit).

In testing such a circuit, if the needle swings appreciably beyond the calculated resistance value (in ohms), the possibility of a short circuit is indicated; if the swing is too low, there may be a poor or loose connection); if the swing is zero, there is an open circuit, or missed connection; if very slight, the very high resistance may indicate leakage to ground (possibly by a bare wire from broken insulation). Any of these indications demonstrates a fault that should be located and corrected before firing the blast.

Fig. 23. Blasting galvanometer.

Open circuits or loose connections can be found in a series circuit by attaching one of the wires to the galvanometer and a probe wire to the other. By progressively probing through the circuit, the weak point can be localized and the fault corrected.

After testing each series cap circuit, the connecting wire circuit is then tested and then the previously tested firing line can be connected in retreat from the face; finally the complete circuit is tested before firing.

The **Blaster's Multimeter** is also furnished with a dry cell, with appropriate resistances and circuitry to limit the test current.

This instrument can be used with a pair of insulated probe wires to check for the presence of stray AC or DC electricity leaking to ground from extraneous sources. These stray currents can be of sufficient magnitude to cause the pemature firing of a cap circuit. They are therefore extremely dangerous. Once located, they should be checked for magnitude. If this indicates a serious hazard, special precautions should be taken to isolate and insulate all parts of the circuit from other obvious conductors, such as rails, pipes and ducts. If a stray current is located in the firing line, it should not be connected to the cap circuit, and blasting operations should not be allowed to proceed until the stray current has been effectively eliminated.

The Blaster's Multimeter is a versatile precision meter designed to measure volts, ohms and milliamperes (i.e., all the electrical factors in a blasting circuit) in one instrument, by selecting the appropriate measuring circuit and scale. It can therefore be used as a superior type of blasting galvanometer (for circuit testing and resistance measuring); as a blasting voltmeter (for voltage and for stray current readings); for blasting machine and power line output voltage readings; and for testing other instruments. With CD type machines, it can be used to give a reading of output voltage which is effective provided the condensers are in good condition. A booklet "All about Atlas Blaster's Multimeter model 616-A" sets out the specifications, operating instructions and other details (January 1975).

A convenient method of indicating continuity in a small wiring circuit is provided by the CIL **Circuit Tester.**[6] This is a pocket instrument designed for circuit resistances up to 75 ohms. It consists of a battery and glow lamp in a small cylindrical housing with metal ends. The lamp glows when the two ends of the unit contact the two wires of a blasting circuit, one at each end (see Fig. 24).

In the interest of safety, all testing instruments should be checked periodically (and calibrated, where necessary) by a competent authority. This procedure is too important to be allowed to provide an opportunity for the misplaced enthusiasm of an amateur.

Before use in underground coal mines, the approval of the MESA should be obtained for any particular make of instrument.

Fig. 24. Circuit tester.

References

1. "Electric Shotfiring Practice", 3rd ed. Schaffler & Co., Vienna.
2. "Blasting Wire Handbook", Seminole Products Inc. Glendora, NJ.
3. *Handbook of Electric Blasting,* Rev. ed., Atlas Powder Company, 1976.
4. Explosives Technical Data Sheet 224C, Hercules Incorporated.
5. "Products for Controlled Blasting," Nitro Nobel, Gyttorp, Sweden, 1974.
6. *Blaster's Handbook, 6th ed.* (Montreal, Canadian Industries Ltd., 1968).

CHAPTER 8

Shotfiring with Detonating Cord

General

Detonating Cord (known variously by such trade names as Primacord, Primex, and Cordeau Détonant) is a cord consisting of a core of initiating explosive (usually PETN) contained within a plastic sheath and wrapped in various combinations of textile, wire and plastic coverings.[1]

It is used for the simultaneous firing of widely distributed charges and for the mass initiation of very large charges, as in quarrying and chamber blasting.

It is safe to handle, extremely water-resistant, and has a high velocity of detonation (21,350 feet per second).

One of the chief advantages of detonating cord (DC) is that, when initiated, it is capable of transmitting the energy of a detonator to all points along its length.

It is reputed to be capable of increasing the efficiency of a blast as it explodes with extreme violence the high explosives (other than blasting agents) lying alongside it in the borehole. It detonates all connecting charges simultaneously from one point of initiation.

In quarry blasting, it is usual to lay out a trunk line of detonating cord extending along each line of holes, connecting to branch or "downlines" into each hole.

With detonating cord, detonators are not required in the holes. Where (as with blasting agents) the charge requires a greater initiating effect than is available from detonating cord, one or more high explosive booster units are attached to the downline (see below).

The whole circuit and series of charges (instantaneous or delay) can be initiated by the use of one No. 6 detonator (plain or electric). The detonator is lashed alongside the trunk line with tape, with its base pointing in the direction of travel of the desired detonation wave.

Alternatively, the trunk line can be initiated

(a)   by a SAF-T-DET Detonating Cord Initiator, a non-explosive electrical device specially designed for the purpose, and marketed by Energy Sciences Corporation of Seattle, WA, or

(b)   by an Exploding Bridge Wire (EBW) detonator RP-80, supplied by Reynolds Industries, Inc. An EBW detonator contains an insensitive explosive charge such as PETN and RDX, fired by a gold bridgewire. Apart from its use as an initiator of DC circuits, it has unique safety features that indicate its preferred use in a number of applications in general blasting operations. Specific firing sets are available for particular applications.

Detonating cord is much safer to handle than a detonator, even though both contain PETN. (It is not the PETN base charge but the priming charge of a cap that renders it so sensitive to heat, impact and friction.) PETN is relatively insensitive.[2] Detonating cord is also safer to use than electric circuits because the whole series of charges can be placed several hours before firing without risk from any hazard including lightning, except perhaps a brush fire. In other words, the whole charge is foolproof until the initiating detonator is taped to the trunk line immediately before firing.[1]

Detonating cord is particularly useful for under-water blasting applications. However, the free ends should be sealed or adequately waterproofed, since PETN will slowly absorb water and become insensitive to initiation. Where the ends have become damp, they should be cut off; or otherwise the detonator should be taped to a dry section.[1,2]

Detonating cord trunk lines may be extended or spliced by regular square knots (Fig. 25). Branch lines should be connected to trunk lines by strapping (taping), by clove hitch, or by a plastic connector. As the detonating wave is uni-directional, it is important that branch lines, when connected by strapping, should have the free end facing the direction from which the impulse is advancing. This precaution is unnecessary when the clove hitch or connector is used.

Plastic connectors cost about two cents each. They are simpler to use, and save much time in connecting branch to trunk lines, especially where the detonating cord has a stiff outer covering (see Fig. 26).

Fig. 25. Square knot in DC trunk line.

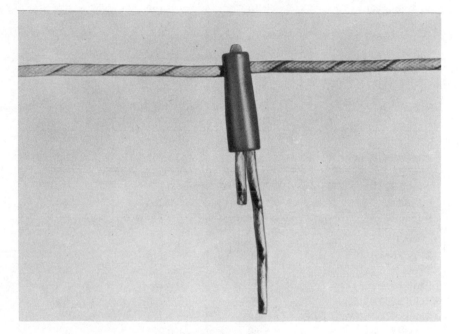

Fig. 26. Plastic connector used to connect DC downline to trunk line.

Detonating cord costs generally about 4 cents per ft. It is available in a number of different loading weights and covering specifications for particular applications (see Table 8). Regular types have a core load ranging from 8 to 60 grains of PETN per ft, with overall diameters of 0.15 to 0.4 inches, depending mainly upon the nature of the encasement. Tensile strengths range from 130 to 375 lb.

TABLE 8
Types of Detonating Cord Available

Trade Name	PETN Core Load (gr/ft)	External Diameter (inches)	Tensile Strength (lb)
AUSTIN:			
Scotch cord	18	0.165	200
A cord	25	0.175	200
40 Reinforced	40	0.190	300
50 Reinforced	50	0.208	200
60 Detonating Cord	60	0.215	325
Tuff-Kote	60	0.240	375
Heavy duty DC	100/150/ 175/200/ 400	0.235/0.312/ 0.330/0.340/ 0.400	275
Seismic DC	50/100/ 150/200	0.195/0.240/ 0.310/0.350	250
CIL:			
Primacord Reinforced	43	0.200	275
Scufflex	54	0.237	300
Trunkline	50	0.176	250
E-Cord	25	0.160	250
B-Line	25	0.130	200
COAST:			
20 gr Primex	20	0.175	170
T line Primex	30	0.150	130
Economy Primex	30	0.185	180
Reinforced 40 Primex	40	0.200	270
Reinforced 50 Primex	50	0.215	270
Tuf-glo Primex	60	0.225	300
ENSIGN BICKFORD:			
Reinforced Primacord	50	0.200	200
Scufflex	55	0.205	300
E Cord	25	0.157	150
Detacord	18	0.150	150

Special types of detonating cord are also manufactured for particular purposes, such as

"Detacord", 18 gr/ft, and "B-line", 25 gr/ft, for secondary blasting;

"Plastic Reinforced Primacord", 54 gr/ft, for under-water blasting;

"PETN 60 Plastic", 60 gr/ft, for oil well servicing;

"Seismic Cord", 100 gr/ft, for seismic work;

"RDX 70 Primacord", 70 gr/ft, for oil well perforating.

References

1. R. A. DICK, "Explosives and Borehole Loading," in SME Mining Engineering Handbook (New York, AIME, 1973).
2. "Primacord Detonating Cord", Ensign Bickford Company, Simsbury, CT. 1976.

CHAPTER 9

Sequential Firing

General

In many applications of the use of explosives, particularly in mines and quarries, provision must be made for certain of the charges to explode in sequence rather than simultaneously. This is accomplished in the following ways:

With Conventional Safety Fuse Firing

(a) By blasting one or two holes at a time and returning at a safe interval after each blast to light the next hole.

(b) By cutting fuses differentially as to length and/or by lighting them in the desired order.

(c) By connecting standard length fuses with IC in the desired order.

With Electric Detonators

(a) By using regular delay detonators.

(b) By using short/millisecond delay detonators.[1]

(c) By using a sequentially controlled timing device with delay detonators.[2]

(1) **Electric detonators (regular delay):** These have a longer shell than the instantaneous electric detonators in order to accommodate a delay element interposed between the fusehead and the priming charge.

There are generally sixteen possible delays, 0 to 15, with a nominal half-second time interval between each sequence number. They are identified by a delay number clearly tagged to the leg wires. The number of the delay is a function of the length of the delay element. Regular delay detonators are used mainly in certain classes of mine development work, and in tunnelling operations, to provide a predetermined sequence of shots and allowing sufficient time for the rock to move from a restricted area. They are generally available in No. 6 and No. 8 strengths, in aluminium tubes.

(2) **Electric detonators (short-delay)**[1]:   These have developed from the regular half-second delay detonator; they represent an important advance in blasting efficiency. The main difference is that the delay interval is much shorter. The delay element contains specially blended powders interposed between the fusehead bridge wire and the priming charge.

Electric delay detonators are packed in cardboard boxes and tagged with the delay period number and the length of leg wires, the ends of which are protected by a shunt (Fig. 27).

These are now supplied by manufacturers in a wide range of delay intervals in two classifications, as either No. 6 or No. 8 strength caps.

(a) **Normal type,**[1] available in a range of 18 to 38 delays, each interval varying from 8 to 100 or more milliseconds (see Table 9). In order to avoid overlap, the interval is increased in the later delays. They are in general use in practically all rock-breaking applications, in copper, gilding metal, bronze, or aluminium shells/tubes. Leg wires range from 4 to 60 ft and more in length. Average price ranges from 40 cents to $ 1.40 each in lots of 5000.

(b) **Non-incendive type,** for use in coal mines, available in a range of up to ten delays, with intervals varying from 25 to 75 milliseconds (see Table 10). These are available in copper shells with copper or iron leg wires from 4 to 16 ft long.

Both types can be used in wet conditions and will withstand water pressures of at least 60 lb/in^2—equivalent to a head of about 140 ft of water.

The fusehead resistance is substantially the same in both types. Although the fuseheads of all detonators in any particular circuit ignite as soon as the current is applied, the detonations do not take place until the respective delay elements have burnt through. Hence, the breaking of the circuit by the first detonation does not affect the firing of the later charges.

However, because of the wide variation of delay intervals in the above types of delay detonators, they should not be mixed in any one blasting circuit; nor should detonators from different manufacturers. Trade names of electric delay detonators are shown in Table 11.

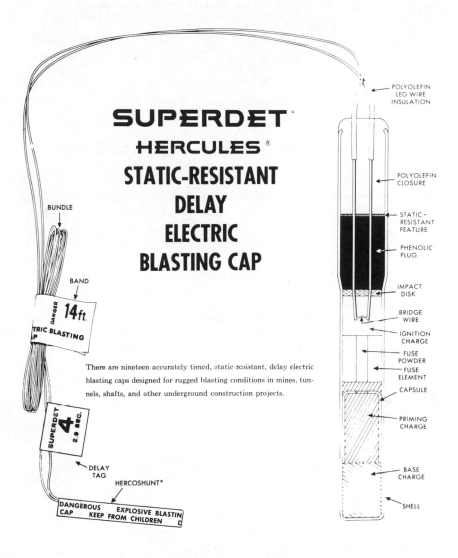

SUPERDET ®
HERCULES ®
STATIC-RESISTANT
DELAY
ELECTRIC
BLASTING CAP

BUNDLE

BAND

DANGER 14ft.

TRIC BLASTING
P

There are nineteen accurately timed, static resistant, delay electric
blasting caps designed for rugged blasting conditions in mines, tunnels, shafts, and other underground construction projects.

SUPERDET 4 2.9 SEC.

DELAY
TAG

HERCOSHUNT®

DANGEROUS   EXPLOSIVE BLASTIN
CAP   KEEP FROM CHILDREN

POLYOLEFIN
LEG WIRE
INSULATION

POLYOLEFIN
CLOSURE

STATIC-
RESISTANT
FEATURE

PHENOLIC
PLUG

IMPACT
DISK

BRIDGE
WIRE

IGNITION
CHARGE

FUSE
POWDER

FUSE
ELEMENT

CAPSULE

PRIMING
CHARGE

BASE
CHARGE

SHELL

Fig. 27. Electric delay detonator.

TABLE 9

Typical Delay Intervals with Short Period Delay Detonators

Delay No.	Atlas (Rockmaster)		CIL (Short Period Caps No. 8)		DuPont (MS Delay)		Hercules (Millidet No. 8)	
	Inter-val msec*	Elap-sed time msec*	Inter-val msec*	Elap-sed time msec*	Inter-val msec*	Elap-sed time msec*	Inter-val msec*	Elap-sed time msec*
0	–	0	–	0	–	0	–	0
1	8	8	8	8	25	25	12	12
2	17	25	22	30	25	50	13	25
3	25	50	20	50	25	75	25	50
4	25	75	25	75	25	100	25	75
5	25	100	25	100	25	125	25	100
6	25	125	30	130	25	150	35	135
7	25	150	30	160	25	175	35	170
8	25	175	30	190	25	200	35	205
9	25	200	40	230	50	250	35	240
10	50	250	50	280	50	300	40	280
11	50	300	60	340	50	350	40	320
12	50	350	70	410	50	400	40	360
13	50	400	80	490	50	450	40	400
14	50	450	80	570	50	500	50	450
15	50	500	80	650	100	600	50	500
16	50	550	75	725	100	700	50	550
17	100	650	75	800	100	900	50	600
18	100	750	75	875	100	1000	100	700
19	125	875	75	950	100	–	–	–
20	125	1000	75	1025	–	–	–	–
21	125	1125	100	1125	–	–	–	–
22	125	1250	100	1225	–	–	–	–
23	125	1375	125	1350	–	–	–	–
24	125	1500	150	1500	–	–	–	–
25	125	1625	175	1675	–	–	–	–
26	125	1750	200	1875	–	–	–	–
27	125	1875	200	2075	–	–	–	–
28	125	2000	225	2300	–	–	–	–
29	125	2125	250	2550	–	–	–	–
30	125	2250	330	2880	–	–	–	–
31	125	2375	230	3050	–	–	–	–
to	125	–	–	–	–	–	–	–
38	125	3250	–	–	–	–	–	–

* msec = milliseconds

TABLE 10

Typical Delay Intervals with Short-Delay Non-Incendive Coal Mine Detonators

Delay No.	Atlas (Kolmaster SF)		DuPont (Coal Mine)		Hercules (Coaldet)	
	Interval (msec)	Elapsed time (msec)	Interval (msec)	Elapsed time (msec)	Interval (msec)	Elapsed time (msec)
0	–	0	–	0	–	0
1	25	25	25	25	25	25
2	75	100	75	100	75	100
3	75	175	75	175	70	170
4	75	250	75	250	70	240
5	50	300	75	325	80	320
6	50	350	75	400	80	400
7	50	400	100	500	50	450
8	50	450	100	600	50	500
9	50	500	100	700	50	550
10	50	550	100	800	50	600
11	–	–	100	900	100	700
12	–	–	100	1000	75	775

TABLE 11

Trade Names of Electric Delay Detonators Supplied by Various Manufacturers
(showing cap number and range of delay periods)

Manufacturer	Regular Delay 300/1500 msec*	Short Delay 25/100 msec*	Coal Mine Applications 25/100 msec*
Atlas	"Time Master SF" No. 6, 1–15 "Rockmaster T" 1–14	"Rockmaster" No. 6, 1–38	"Kolmaster" No. 6, 1–10
CIL	"LPV Delay" No. 8, 1–15 "Long Delay" No. 8, 1–20	"Short Period Caps" No. 8, 1–30	"Non-Incendive Short Period" No. 8, 1–10
DuPont	"Acudet Mk V" 1–14	"MS Delay" 1–19 "Strip-Det Delay" 1–10	"Coal Mine Delay" 1–12
Hercules	"Superdet" No. 8, 1–15	"Millidet" No. 8, 1–17	"Coaldet" No. 8, 1–12

*) milliseconds

(3) **Sequentially controlled timing devices:** One of the main problems in delay firing of large connected segments in the one firing by power mains is the need to balance the resistances in each segment effectively; and to control air concussion and ground vibration, as well as to improve fragmentation.

An automatic solid state *sequential timing device* has been developed by Research Energy of Ohio. This unit produces accurate timed pulses of energy for detonating blasting caps.[2] The elapsed time interval can be adjusted from 25 to 250 milliseconds.

The BM-75 Sequential Blasting Machine is a similar device, marketed by Austin. Trojan supplies the Trojan C Timer.

With Detonating Cord or other Non-electric Delay Systems

(a) By using non-electric milli-second (M/S) delay connectors in the trunk lines.

(b) By using an all-electric trunk line circuit with delay detonators taped to DC downlines; or

(c) By using an integral non-electric delay system.

(1) **Non-electric M/S delay connectors:** These are provided in copper tubes about three inches long with an explosive charge in each end, separated by a central spacer element that incorporates the delay mechanism. Detonating cord tail pieces are factory-crimped to both ends to connect with square knots to the gap cut in the trunk line (Fig. 28). As supplied by Austin, there are MS-9, MS-17 and MS-25 millisecond delays available, with provision for MS-35 and MS-45 for special purposes .Each delay period is identified by color.

Austin also produces a non-electric *ADP Primer and Relay* as a standard unit incorporating a cast booster within a plastic housing, and which can be simply armed with one of a range of eleven *M/S Delay Relays,* any number of which can be conveniently connected to a single 25 gr downline.[3] These relays are designed for 25/50/75/100/125/150/175/200/250/300 and 350 millisecond delays; and there are special relays for 5/9/17/35 and 45 millisecond delays. Each booster unit is made with special cord channels for ease of assembly.

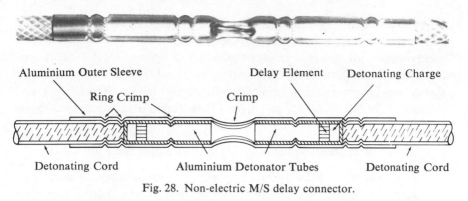

Fig. 28. Non-electric M/S delay connector.

DuPont provide a different style of "MS Connector" for surface firing, to be set in a pre-cut gap in a detonating cord trunk line system. These consist of a moulded plastic sleeve containing a copper tube delay element in the centre. Each end of the sleeve is formed in such a way that a free end of the primacord gap can be looped and locked in a slot with a tapered sawtooth pin in such a position that the cord makes physical contact with the copper tube delay element. In this way, a safe foolproof non-electric method of delaying a trunk line can be conveniently provided. Delays of 5, 9, 17, 25, 35 and 45 milliseconds are available, each in a different identifying color.

As with the Austin Delay Connectors, an almost infinite array of delayed firing patterns can be provided by combining connectors of different delay periods.

For surface mining, delays can also be effected with a NONEL Primadet inserted in a NONEL shock tube type of trunk line (see later). Precise delay timing is secured with the additional advantage of noiseless operation. These Noiseless Trunkline Delays are available from Ensign Bickford in 5, 7, 8, 9, 17, 25 and 42 milliseconds in lengths of 20, 30, 40, 50 and 60 feet.

(2) **Integral Non-electric Delay Systems:** These involve a method of introducing sequential firing mechanisms into non-electric circuits, while retaining the advantages of precise short delay blasting available in electric circuits. In this way, the relatively unsafe features of electric blasting in surface operations (due to lightning, radio-frequency energy, and static electricity) can be avoided. These systems are particularly applicable to bench holes charged with insensitive material such as AN/FO, SBA and SE. Several such systems have recently been developed by different manufacturers.

Primadet delay system: Primadet Delays were developed by Ensign-Bickford to provide a practical non-electric detonation system that would yield precise timing characteristics equivalent to those of electric delay systems, free from static and stray current hazards.[4] The delay caps of this system are incorporated in primers within the borehole charge. They are available in both short delay and regular half-second delay series (see Table 12).

The Primadet Delay System consists of three major elements.

(1) A special No. 6 blasting cap, incorporating its own inbuilt delay element as an integral feature. It should be inserted into a cartridge or cast booster of adequate strength and sensitivity.

(2) A miniature 4 gr/ft PETN detonating cord called *Primaline,* one end of which is factory-crimped into the delay cap, as a "downline", in standard lengths ranging from 6 to 50 ft.

(3) A plastic "J-connector" for readily attaching the free end of the Primaline to the trunk line. The J-connector is color-coded blue (for use with orange Primaline in the short delay system) and bears the number of the delay period; or red (with yellow Primaline) for the regular half-second delay system (see Table 12).

It is usual to initiate the Primaline downline (which takes the place of leg wires or safety fuse in other detonating systems) by a trunk line of PD cord or Reinforced Primacord. It is important to avoid using the Primaline as a downline for sensitive explosive charges such as NG-based types, since it may initiate such a charge, perhaps imperfectly, and cause it to detonate instantaneously rather than in delayed order.

Similarly, the type of primer cartridge through which the Primaline is threaded or laced, must be insensitive to initiation from its action, yet of sufficient strength and sensitivity to detonate the main charge at the proper instant.

Primaline downline is now used mainly in connection with HD NONEL Primadets for surface mining applications. When Primadets (LP and MS series) are used in underground mining applications, they are now connected with a NONEL shock tube instead of Primaline.

Primadet delays can therefore be used effectively only in charges of low initiation sensitivity such as blasting agents (AN/FO types or slurries). They are packed as self-contained "package units" of all three components and labelled according to their delay type (MS or LP) and the length of Primaline lead.

In Canada, CIL provides a **"Toe-Det" Delay System**,[5] using a similar special delay cap linked permanently to a reinforced **Anoline** low strength DC downline which is in turn connected by a J-shaped plastic connector to a specially designed high strength **Trunkline** detonating cord. It is essentially a short-delay system with 30 delay periods ranging from 25 to 250 milliseconds. Specially developed Pro-core type cast **Toe-Det boosters** have been produced to accommodate the Toe-Det delay assembly.

The Anodet Delay Blasting System[6] is also produced by CIL. It is particularly designed for underground blastholes, 1 to $2^{1}/_{4}$ inches in diameter, pneumatically loaded with AN/FO. The Anodet system consists of a special

TABLE 12
NONEL Primadet Delay Intervals

Delay period	MS delays		LP delays*	
	Interval (msec)	Elapsed time (msec)	Interval (msec)	Elapsed time (sec)
0	–	0	–	–
1	25	25	–	0.2
2	25	50	0.2	0.4
3	25	75	0.2	0.6
4	25	100	0.4	1.0
5	25	125	0.4	1.4
6	25	150	0.4	1.8
7	25	175	0.6	2.4
8	25	200	0.6	3.0
9	50	250	0.8	3.8
10	50	300	0.8	4.6
11	50	350	0.9	5.5
12	50	400	0.9	6.4
13	50	450	1.0	7.4
14	50	500	1.1	8.5
15	75	575	1.1	9.6
16	75	650	–	–

* LP = Long Period (regular half-second delays)

high strength detonator, factory-crimped to a length of Anoline, which is connected to the DC trunkline by a J-connector. Where necessary, an A-3 or Anodet Booster can be used. To locate the primer centrally in the blast-hole, *a Plastic Cap Holder* is available. Anodet delays are available in short or long periods. Anodet Short Delays are supplied in a range of 30 delays with intervals from 20 to 250 milliseconds. Anodet Long Delays range from 275 to 1050 milliseconds over 20 delay periods. The Anoline downlines are supplied in 3, 4 or 5 metre lengths.

The NONEL system: This Swedish invention by Nitro Nobel AB is a nonelectric system that does not involve the use of detonating cord.[7] Four components are included in a complete unit, supplied in a standard pack. These are:

(a)   a NONEL tube:

(b)   a conventional (plain) detonator with a delay element;

(c)   a connecting block, provided with a minidetonator;

(d)   a starting gun and NONEL trunk line.

The NONEL tube, of appropriate length, is connected at one end to the delay detonator. The other end is connected to the yellow *connecting block* that houses the *minidetonator* (or transmitter cap). The transparent plastic NONEL tube is 3 mm in external diameter with a 1.5 mm bore. The inside wall of the tube is coated with a low concentration of explosive powder that possesses the ability to conduct a shock wave at constant velocity. The shock wave is supplied by the minidetonator in the connecting block. It travels through the tube and emerges in the detonator as an intensive tongue of flame. The complete circuit is initiated by a special starting gun that ener-gizes a NONEL trunkline which in turn initiates each connecting block connected to it. The NONEL detonators are supplied in a range of 20 delay intervals each of 25 milliseconds, and six more each of 100 and 150 milli-seconds.

The Combination NONEL Primadet system: Ensign have combined the advantages of these two systems, and now supply this combination system for a variety of precise non-electric initiation hook-ups (LP and MS) for underground mining applications.

The Hercudet Blasting Cap System also eliminates all the hazards asso-ciated with the use of electric detonators.[8] Unlike DC initiation, it is also virtually noiseless. It consists of three major elements.

(1) A special aluminium shell Hercudet detonator, incorporating a delay element, and furnished in the factory with two plastic tubes, rather than with two leg wires (see Fig. 29).

(2) Hercudet connectors for connecting lengths of tubing between adjacent holes in the circuit. Tee connectors are also available.

(3) Hercudet Blasting Machine (with Bottle box and Tester).

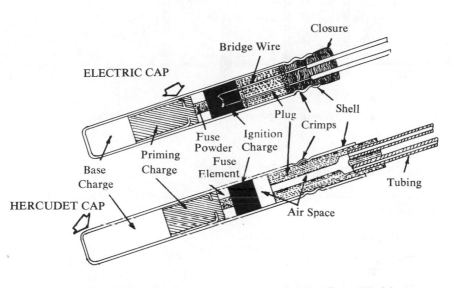

Fig. 29. Hercudet detonator, as compared with ordinary EB delay cap

After the detonators (in their primers) have been placed and the charging of the holes completed, the tubing downlines are connected in the circuit to the trunkline (with connectors), and to the blasting machine (see Fig. 30).

To this stage, the circuit tubing has contained nitrogen. The circuit may now be purged with air and tested. To fire the round, the valves on the bottle box are opened to charge the blasting machine with the firing mixture of fuel and oxidizer, the "arm" button is pressed for a short time, and then the "fire" button. This initiates the gas mixture in the ignition chamber of the blasting machine, resulting in a detonation that proceeds around the tube circuit at about 10,000 ft per second, and initiates all the detonators.

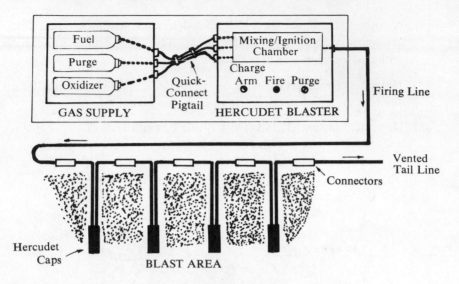

Fig. 30. Schematic Representation of Hercudet System Circuit.

Advantages of Short-Delay Blasting

The advantages of short-delay over instantaneous or half-second delay blasting are:

1. Reduction of ground vibration.

2. Reduction in air concussion.

3. Reduction in overbreak.

4. Improved fragmentation.

5. Better control of fly-rock.

This method derives its advantages from the fact that, in the shorter time-interval involved, while the rock mass is still under the influence of the shock wave from the preceding detonation, the next (and, in turn, each succeeding) detonation occurs.[9]

Thus the superimposing of each new detonation wave on an already strained rock mass more readily overcomes its cohesive strength, in accordance with what is known as the Principle of Mutual Assistance. This results in the advantages listed above, for the same weight of charge, or alternatively, the need for a smaller charge.

One of the most important advantages noted with short-delay blasting is the marked reduction in vibration as compared with that obtained when an equal weight of explosive is detonated instantaneously. Vibrograph recordings indicate that the effective vibration bears a close relation to the weight of explosive detonated on any short-delay period, rather than the total used in a blast; and in practice it would seem that if blasting can be conducted with instantaneous initiation on a given site without causing severe vibration when using, for example, 120 lb of explosives, the use of 10 delay periods of the order of 25 milliseconds will permit a single blast of a total weight of 1,200 lb. Experience has shown that on the majority of sites **two-thirds** of the instantaneous charge limit can be fired on each of the short delay periods without giving any increase in ground vibration. Considering the above example it should be safe, under normal conditions, to fire as many 80 lb delay charges as there are shortdelay intervals available. It is always advisable when vibration is critical to check the site by means of vibrograph recordings from typical blasts. Where problems are much less acute, but where the operator feels that it is in his best interest to reduce the ground vibration from his blasts, short-delay blasting methods can be introduced with the certain knowledge that they will assist to this end.

Structural Damage Criteria

This question refers to damage that may be caused to structures such as residences, public and commercial buildings, churches, bridges, towers and monuments situated near a blast site, chiefly due to the effects of ground vibration.

The increasing number of damage suits filed against quarry and construction companies in recent years demonstrates the need for improved ground vibration technology.

As stated by MILLER:[10]

"Blasting vibration and air [concussion] are two conditions that can cause people living in the area of a blast site to register complaints — not only to the operator doing the blasting but also to governmental agencies who may exercise some control over the operator's activities. Severe vibration and air [concussion] can cause structural damage to nearby buildings; these are the conditions that need to be controlled. On the other hand, less severe vibration and air [concussion] that are incapable of causing structural damage can still be felt or heard by people. The reaction to these low levels of vibration and noise will depend on the subjective response of the particular individuals. It could range from no reaction, to annoyance or even to claims of damage to their homes.

It is necessary to control vibration and air [concussion] both to prevent damage to nearby structures and to minimize complaints."

It is convenient to analyze a ground vibration, as it affects a particular structure, in three orthogonal components of motion: horizontal longitudinal, horizontal transverse, and vertical. In any given case, the component that exhibits the maximum effect is the important one. In most cases, it is the longitudinal horizontal component, in the direction of the blast site.

Early work in this field was conducted by the U.S. Bureau of Mines between 1935 and 1942, leading to the issue of Bulletin 442 in 1942.[11] More recent work, as outlined in Bulletin 656, issued in 1971, points to the following conclusions:[12]

1. Particle velocity is the best criterion for predicting the probability of vibration damage to structures.

2. Particle velocities less than 2 in/sec show little probability of causing structural damage while particle velocities greater than 2 in/sec are more likely to cause structural damage.

3. If there are at least eight milliseconds (0.008 sec) between detonations, the vibration effects of individual explosions are not cumulative.

These conclusions led to a simple equation for predicting a safe distance for a given weight of explosives to avoid structural damage. The following relationship was set up.

$$v = H \left(d \sqrt{E} \right)^{-\beta}$$

where

v is the particle velocity, inches per sec

d is the distance from the shot point to the structure, ft

E is the maximum weight of charge per delay interval, lb

H is a constant for the particular ground conditions

β is an exponent determined from test results.

However, to use this equation intelligently, it is necessary to have values for H and β.

These can be determined experimentally by using a seismograph at the site. But otherwise they can be avoided by using the Scaled Distance Formula, as set out below.

The expression $d/\sqrt{E}$ is referred to as a "scaled distance" in ft per square root of charge weight in lb. From plots of site experimental data, a scaled distance can be selected for which a corresponding safe vibration level can be assured. Once a safe minimum scaled distance has been so determined (usually in excess of 50 ft per square root of charge weight, lb), the safe charge weight per delay, E_s, for any blast can be determined by:

$$E_s = (d/D_s)^2$$

where

d is the distance between the blast site and the structure, ft; and

D_s is the minimum safe scaled distance allowable.

Where a value of D_s of 50 or more is used, then the maximum charge weight E_s per delay can be determined safely without the need of instrumental trials. However, if larger charge weights are required, then instrumental measurement should be undertaken to check whether or not a safe level of vibration is likely to be exceeded.[12, 13]

Sophisticated instruments for measuring components of particle velocity in ground vibrations are now available.[14]

Individual state authorities have adopted varying regulatory methods and values for enforcing safe vibration limits in built-up areas.

References

1. *Handbook of Electric Blasting*, Atlas Powder Company, Rev. Edn., 1976.
2. N. P. CHIRONIS, "New Blasting Machine Permits Custom-programmed Blast Patterns", *Coal Age*, March 1974.
3. Austin Technical Data Bulletin DP477.
4. "Primadet Delays", Technical Bulletin EB5, Ensign Bickford Company, 1969.
5. "Toe-Det Delay System", Information Report No. 140, Canadian Industries Ltd.
6. "Anodet Delays", Information Report No. 115A, Canadian Industries Ltd.
7. "Nitro Nobel Introduces the NONEL Detonator", Nitro Nobel AB.
8. R. B. HOPLER, "The Hercudet Non-electric System of Blasting", Paper presented to Annual Meeting of Society of Explosives Engineers, Pittsburgh, Feb. 1977.
9. M. A. COOK, *The Science of High Explosives*, (New York, Reinhold, 1958).
10. P. H. MILLER, "Blasting Vibration and Air Blast", Atlas Powder Company.
11. J. R. THOENEN and S. L. WINDES, "Seismic Effects of Quarry Blasting", *U.S.B.M., Bull. 442*, 1942.
12. H. R. NICHOLLS, C. F. JOHNSON, and W. I. DUVALL, "Blasting Vibrations and their Effects on Structures", *U.S.B.M. Bulletin*, 656, 1971.
13. J. F. DEVINE, R. H. BECK, A. V. C. MEYER and W. I. DUVALL, "Effect of Charge Weight on Vibration Levels from Quarry Blasting", *U.S.B.M., R.I. 6774*, 1966.
14. G. A. BOLLINGER, *Blast Vibration Analysis*, (Carbondale, IL, Southern Illinois University Press, 1971).

CHAPTER 10

Special Forms of Explosives

Liquid Oxygen Explosive (LOX)

A liquid-oxygen explosive[1] is a detonable mixture employing liquid oxygen as the oxidant of a combustible absorbent material . . . The detonation is the result of the rapid reaction of the combustible with the oxygen, which can occur over a wide range of oxygen concentration. A large variety of ingredients has been used in liquid-oxygen explosives.

Liquid-oxygen explosives are unique in that the oxidising component of the explosives system is not permanently held, as in the case of fired explosives.

. . . upon exposure to surroundings at ordinary temperatures, evaporation . . . of the liquid (oxygen) proceeds at a rate determined by the rate of heat transfer to the liquid.

If evaporation of the liquid oxygen is permitted to continue the cartridge loses its excess oxygen; and eventually the oxygen content falls below the amount required for good explosive performance, ultimately becoming non-explosive.

Combustible materials used include lampblack, carbon black, or char. Cartridges of this material, dipped in a vacuum-walled container of liquid oxygen, are charged into a blast hole and fired with a detonator. The explosive effect is produced by the rapid reaction between the oxygen and the combustible, after detonation; this forms a large volume of CO_2 gas at high temperature and pressure.

The charge should be fired as soon as possible, as the potential energy of the explosion falls off with time as the evaporation of the liquid oxygen proceeds.

This type of explosive presents unique safety features. Apart from the risk of accidental fire, the explosive is basically safe after the liquid oxygen evaporates.

The economic use of LOX depends largely on the availability of a cheap source of oxygen. Since the recent rapid development of blasting agents, it seems unlikely that LOX will ever become a widely used explosive. About 1000 tons is used annually for stripping overburden in the United States.

Shaped Charges [2, 3, 4, 5, 6]

The application of shaped explosive charges to mining operations has extremely interesting possibilities in the field of cost reduction in the drilling of shot-holes.

The principle, known as the "Munroe effect", involves the concentration of the force of the explosion in the form of a jet. The principle was used during World War II for demolition purposes, for certain grenades, and for the "bazooka" which employed a warhead containing a shaped charge lined with a metal cone.

Some non-military applications include the "explosive rivet" (used in aircraft construction), the "jet perforator" (for perforating oil-well casing), and the "jet tapper" (for tapping blast funaces).

Possible mining applications include the drilling of blastholes or the blasting of an entire cut in a development face.

A charge described by AUSTIN[4] consists of four components: a detonator, 3 · 3 lb of high explosive, a 4 inch diameter cylindro-conical body, and a metal or ceramic liner of 55° included angle. The charge is placed with the conical liner about 7 inch from the rock face. On detonation, the liner collapses and is expelled as a jet with a forward velocity of up to 30,000 ft/sec. The impact of the jet disintegrates the rock target, forming a hole up to 3 inch diameter and up to 5 ft deep. As at present developed, great care is required in its use. Apart from a considerable air-blast effect, there is a danger from projection of rock fragments, especially with imperfectly directed shots.

In other fields, the development of shaped charges has been markedly more successful,[5] particularly where conventional explosives are difficult to use as, for example, in under-water applications. One of the pioneer operators in this field is Jet Research Center, Inc., of Arlington, TX (see Chapter 17).

Goex has developed a shaped charge system for breaking boulders (see Chapter 14).

Substitutes for Explosives [6, 7]

The main field of use for commercial explosives is in breaking hard rock: to minimize the arduous nature of the work, and to increase the productivity of such operations (see Chapter 1).

The universal employment of explosives for such purposes has, for a century or more, established the place of explosives as an important tool of engineering endeavor.

However, since the development of hard metals (tungsten carbide) in World War II, and of other techniques, excavation of the less tenacious rocks is being increasingly accomplished by non-explosive methods. Some of these are set out below.

1. **Capsuled compressed gas cartridges:**[6, 7, 8, 9, 10] Two general types of non-explosive compressed gas systems have been developed primarily for "soft ground" applications.

These are used in underground coal mining (see Chapter 20), for loosening compacted bulk materials in storage, and for dislodging accretions from hot furnaces. In all of these applications, safety is enhanced. They depend for their action upon the build-up of gas pressure in a cylinder, which, following a trigger action, is suddenly expanded into the material to be broken or loosened. With coal mining, the cylinders are located in regular boreholes. For loosening bulk materials, or breaking up furnace accretions, the cylinder is inserted into similar holes; or (for Airdox), located permanently in a recess in the wall of a storage bin.

For coal mining, a mobile machine is available to drill holes and place cartridges of **Airdox;** or a specially designed self-propelled "shooting car" is used for multiple sequence shooting.

Airdox cylinders are inserted in drill holes in the face and charged to 10,000 to 12,000 lb/in^2 by a nearby high pressure six-stage compressor through a feeder hose. Cylinders (which are re-usable) are fired remotely through shooting hoses. This equipment is available from the manufacturers on lease arrangements.

For loosening bulk materials, **Cardox** cylinders are also available; but especially for intermittent *ad hoc* duty at infrequent intervals. This unit consists of an alloy steel cartridge filled with liquid carbon dioxide and a chemical igniting agent. When required, the reaction is initiated by an electric current controlled from a safe distance. The expanding action of the CO_2 gas breaks a soft steel disc at the discharge port, and the escaping expanding gas "pressurizes" and heaves the enclosing material. Cardox cylinders may be re-charged at the factory. Standard discharge pressures range from 10,000 to 19,000 lb/in^2, depending upon the application.

Where there is a continuing routine need for loosening stored bulk materials, the Airdox system is recommended.

Both systems are available from the Long-Airdox Company, Oak Hill, West Virginia.

2. **Mechanical methods:** Tungsten carbide tipped steel (rotating and traversing) abrading elements are now used to excavate rock of soft to medium hardness.[11]

Specialized "continuous miners", tunnelling moles, roadheaders, and large diameter shaft/raise borers have been developed recently to employ these principles in soft to medium rock without the use of explosives. However, for hard rock, where the compressive strength generally exceeds about 35,000 lb/in^2, and where such mechanical methods are therefore uneconomical, explosives must still be used in the conventional drill/blast/muck cycle of operations.[11, 12, 13]

Secondary breaking of large hard rock boulders by thermo-mechanical methods has recently been under investigation.[13] The rock is first heated and then fragmented by mechanical impact.

3. **Electrical methods:** Various methods of breaking rock by electrical energy have been investigated, including heating by induction, high frequency, microwave, electron beam, plasma jet, arc, and condensor discharge techniques.

The "Electrofrac" method uses electrode contact and resistance heating to stress the rock and to create fractures for subsequent separation by mechanical handling. Field trials on electrical fragmentation for secondary breaking of magnetic iron ore have been demonstrated. It is claimed that it is feasible to use this method for primary breaking of iron ore.[14]

4. **Liquid jet systems:** Pulsed water jets at pressures of 100,000 lb/in^2 can fracture rock and concrete. By incorporating a battery of high pressure water jets on a tunnelling machine, it is believed that conventional tunnelling practices will be revolutionized.[12, 15, 16, 17]

5. **"Non-explosive explosives":** Two new forms of binary explosives have been introduced:

(1) **Astro-Pak** is being produced by EXCOA of Issaquah, WA, and is represented as a high-performance 2-component liquid explosive that is safe when unmixed, but can develop a very high explosive power when mixed and placed.[18]

(2) Atlas is marketing a range of binary explosives consisting of an inert solid and an inert liquid, each of which can be shipped and stored as a non-explosive.[19] When the two components are mixed and charged, they form an extremely high strength capsensitive explosive.

(a) **KineStik** is designed for use as a two-component booster or main charge in drill holes as plastic cartridges $1^1/_4 \times 7$, 2×9, $1^1/_2 \times 8$, which when mixed can be initiated with plain or electric detonators or 50 gr DC, yielding a VOD of 18,000 ft per second.

(b) **KinePouch** is designed primarily for secondary blasting. The dry solid component is packaged in a pliable plastic "pouch", to which the liquid component can be added, and the detonator then placed. The pliable pouch can be made to conform to the surface of the boulder to be blasted.

(c) **KinePak** is a ready-mixed capsensitive type of cast booster, $2^3/_4$ inch diameter $\times$ 5 inch long, which can be armed with a Primadet or DC, for maximizing the initiation of AN/FO or slurries. It can develop a VOD of 20,000 ft per second.

References

1. "Safety and Performance Characteristics of Liquid-Oxygen Explosives", *U.S. Bureau of Mines Bulletin,* No. 472 (1941), pp. 1—2.

2. JOHN B. HUTTL, "The Shaped Charge for Cheaper Mine Blasting", *Engineering & Mining Journal,* May 1946, pp. 58—63.

3. R. S. LEWIS and G. B. CLARK, "Application of Shaped Explosive Charges to Mining Operations", *Bulletin* No. 1, *Dept. of Mining Engineering, University of Utah* (July 1946).

4. C. F. AUSTIN, *Min. Cong. Jnl.,* Vol. 50, No. 7.

5. P. De FRANK and C. H. BROWN, "Underwater Explosives Technology", *Marine Technology Society,* 6th Annual Reprints, Vol. 1, June 1970.

6. MELVIN A. COOK, *The Science of High Explosives* (New York, Reinhold, 1958).

7. R. S. LEWIS and G. B. CLARK, *Elements of Mining,* 3rd. ed. (New York, Wiley, 1964).

8. G. W. DAVIS, "Pneumatic System Dislodges Bulk Material Safely and Effectively", *Coal Age,* October 1970.

9. G. W. DAVIS, "Safely Dislodging Material Stoppages", *Compressed Air Magazine,* March 1971.

10. Airdox Shooting. Bulletin 63-1-5000. Long-Airdox Co.

11. I. R. MUIRHEAD and L. G. GLOSSOP, "Hard Rock Tunnelling Machines", *Trans. Instn Min. Metall. (Sect. A: Min. Industry),* 77, 1968.

12. P. J. G. du TOIT and L. P. COLVIN, "Underground Mining Advances with New Breaking and Handling Systems", *World Mining,* Catalog Survey Directory, June 25, 1971, p. 81.

13. K. THIRUMALAI, S. G. DEMOU and R. L. FISCHER, "Thermomechanical Method for Secondary Breakage of Hard Rocks", *U.S.B.M., R. I. 8057,* 1975.

14. E. SARAPUU, "Electrical Fragmentation of Magnetic Iron Ores." Paper presented at AIME Annual Meeting, Washington, DC. February 1969.

15. Water Jet Technology. Rockville, MD; Exotech Inc., Hydromechanics Division.

16. J. N. FRANK and J. W. CHESTER, "Fragmentation of Concrete with Hydraulic Jets", *U.S.B.M., R.I. 1572,* 1971.

17. M. M. SINGH and P. J. HUCK, "Rock Breakage by High Velocity Water Jets", Paper presented to SME Meeting, St. Louis, Oct. 1970.

18. S. W. HURLBUT and B. A. KENNEDY, "Open Pit Equipment Continues Growth and Shrinks Cost per ton handled", *World Mining,* Catalog Survey Directory, June 25, 1971, p. 61.

19. Blasting Data Sheets 102, 103, 104. Atlas Powder Company.

CHAPTER 11

Practical Usage of Explosives

Placement of Charges

The bulk of the explosives marketed in North America is used for "breaking ground" in mines, collieries, quarries, and rock excavations. In order to obtain the most effective use of explosives, they are confined in boreholes drilled in strategic positions in the rock (as discussed in Part II of this textbook).

The effect of the detonation is to weaken the cohesive strength of elastic rocks by *brisance,* reinforced by the high gas pressure created in the charge hole. This detonation sets up in the rock a compressive strain pulse which travels in all directions from the charge hole and attenuates to zero, unless and until it reaches a free face (see Chapter 12), in which case it is reflected as a tensile strain pulse. The rock is therefore broken in tension, aided and displaced by the high gas pressure (bubble energy).

The operations involved in charging a normal short hole with conventional cartridged explosives are described as follows:

1. The hole is first cleaned out with compressed air by means of a blow-pipe.

2. The cartridges of explosive are inserted into the borehole one by one and firmly squeezed into position (so as to occupy the full cross-section of the hole) by means of a wooden tamping pole.

3. The primer cartridge (if of conventional safety fuse/plain detonator type) is placed last.

4. The hole is sealed by stemming with a cartridge of sandy clay material squeezed against the primer.

5. Care is taken to see that no damage is caused to the safety fuse, the free end of which is now protruding from the mouth of the hole.

6. If the charge is to be fired electrically, then the electric primer is the first cartridge to be inserted (inversely) in the hole. The leg wires are then held taut to one side of the hole while the ordinary cartridges are tamped into position.

7. With either type of primer the live end of the detonator should face the bulk of the charge, i.e., with the ordinary primer—inwards, with the electric primer—outwards.

8. On no account must a metal tamping rod (e. g., drill steel) be used.

9. The primer cartridge itself should not be tamped or squeezed with the tamping pole.

When charging soft plastic cartridges of slurries, adjusted techniques need to be adopted. Pneumatically operated cartridge chargers are manufactured by Nitro Nobel AB and supplied by VME-Nitro Consult Inc. They are available in a range of sizes and designs for charging boreholes up to 4 inches in diameter for either rigid dynamite or plastic slurry cartridges. DuPont recommends this device for charging small diameter cartridged slurries such as Tovex. One type can be used in conjunction with a Robot Tamping Machine.

Where the rock formation is of such a nature that it tends to break back and "cut off" part of a neighboring hole that is timed to explode later, the holes should be "bottom-primed", i. e., the primer cartridge should be loaded first, in the bottom of the hole. This can be done safely and effectively *only with an electric primer* (or detonating cord, for deep holes).

There are several reasons why a plain (safety fuse/cap) primer should be used only for top (collar) priming, i.e ., placed last in the hole, immediately before the stemming.

(1) If placed at the bottom of the hole, a sharp reverse 180° bend needs to be made in the fuse (see Fig. 31). This stresses the textile wrappings, thereby permitting water to enter, or the flame to spit outwards, with the possibility of setting fire to the main charge. Admittedly, some manufacturers supply a special fuse with extruded plastic jacket to obviate this problem; but a miner trained to use indirect (reverse-end, bottom) fuse priming will always thereafter tend to do so with whatever fuse is available at the site.

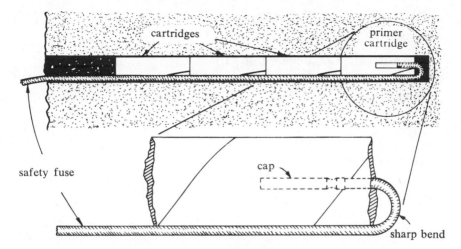

Fig. 31. Incorrect method of bottom-priming (with safety fuse/plain detonator type primer)

(2) Where a bottom-primed hole is well-stemmed (and this is a very necessary requirement for effective blasting), the burning speed of the fuse, under the resulting conditions of superior confinement, can increase by over 60 per cent, in some cases burning almost instantaneously.[1,2] In such a case, a premature explosion may result, yielding a fatal accident due to a "fast fuse".

(3) In other circumstances, the fuse, lying parallel to the main charge, may "side-spit" and ignite the charge before the main flame reaches the detonator, thereby causing a premature explosion (or at least, an unsatisfactory blast).

(4) Again, the heat generated by the burning of the powder train may be transferred laterally to the main charge, raising its temperature to considerably more than 150° F, with risk of a premature explosion. This can occur even without the side-spitting action mentioned under (3) above.

Charging of small diameter underground blastholes with AN/FO (in any direction) is usually done pneumatically. There are two basic types of pneumatic loading machines:

1. **Pressure type,** in which 5—40 lb/in² compressed air is applied above the AN/FO contained in a pressure vessel, causing it to flow from the bottom of the vessel into the placement hose. It is not suitable for charging long "up" holes (unless provided with suitable accessories) because the

particle velocity is usually insufficient to cause the material to stick in the hole. This type of charger is represented by:

ANOL Loader (VME-Nitro Consult)

Powder Monkey (Gulf)

CIL Blasthole Charger

CIL Satellite Loader

These units are generally used for large faces. They have high loading rates, and generally load prilled AN/FO to a density of about 0.8 to 0.9 gm/cc.

2. **Ejector type,** in which the AN/FO is drawn from the bottom of an open-topped container and ejected into the air stream in the placement hose which conveys it into the borehole under a pressure of 65—100 lb/in^2. Typical loaders of this type are:

Gulf Porto-placer

Penberthy Anoloder (CIL)

PORTANOL (VME-Nitro Consult)

The ejector type generally loads to a higher density. It is therefore more suitable for charging long "up" holes, but it operates at a much slower rate. Both types are made in various sizes and capacities.

Combination methods are also used to develop the advantages of both systems. In such an arrangement, the pressure unit and the ejector operate in series (see Fig. 32).

Typical of the combination (pressure — ejector) models are the Atlas Tetloder, the CIL Model 60R Intermediate Loader, and the JET-ANOL Loader (VME-Nitro Consult).

When AN/FO is charged pneumatically, **static electricity** may be generated as the mixture is being blown into the borehole; this can be especially hazardous when holes are bottom-primed with electric detonators in rock of low conductivity under conditions of low relative humidity.[3] In such circumstances, it is safer to "collar-prime" (see Glossary). Otherwise, and particularly for bottom priming, a set of safe requirements has been drawn up by Litchfield, Hay & Monroe, as follow.[4]

Fig. 32. Combination pressure-ejector
type pneumatic blasthole charger.

(1) A semi-conductive charging hose should be used.

(2) The resistivity of the AN/FO should be controlled, either by dampening the AN/FO charge itself, or the walls of the hole.

(3) The borehole should be discharged prior to hook-up of the detonator leg wires.

(4) The operator should discharge himself prior to handling the detonator leg wires.

(5) Tests should be made with a suitable galvanometer to assure the continuity of the detonator wire circuit prior to inserting the EB cap in the hole.

(6) The leg wires should be kept shunted, but not otherwise connected to ground during loading.

(7) The leg wires should never be allowed to contact the loading machine.

Use of AN/FO Underground

AN/FO can be used successfully underground except under wet conditions. For this reason, it is not generally used for shaftsinking, winzing or in other wet blastholes.

Properly mixed and primed charges have good fume characteristics and do not cause headaches when handling.

When charging by pneumatic methods, hole space is more effectively filled and the charging operation is faster than with cartridged explosives. Fragmentation is generally improved, probably because of the better coupling and greater gas volume. AN/FO has been successfully used for small diameter holes for both short and long (ring-drilled) holes.

For primary breaking of ore in stopes, bagged AN/FO is taken by supply cars to feed pressure-type pneumatic loaders located on main levels (in the case of sub-level open stopes), on sub-levels (in the case of sub-level caving), or actually in the stopes (for shrinkage or cut-and-fill stoping).

For secondary breaking, bagged AN/FO is supplied to scram drifts or grizzly levels. It is then transferred in small polyethylene bags for plastering, "browfiring", or chute blasting.

When blasting in development headings, bagged AN/FO is taken to the face in the normal explosives carrying bags. It is pneumatically loaded into blastholes through $5/8$ and $3/4$ inch diameter semi-conductive hoses by ejector-type loaders. Holes are bottom-primed electrically under conditions calculated to avoid static currents (see above). If collar-primed (by safety fuse/caps), igniter cord is used; this may be ignited from a remote place by electric igniters.

Mixing of Blasting Agents

Following some years of experimentation, AN/FO is now usually mixed on the job site in the following ways. For small-scale operations, the correct amount of fuel oil is introduced into a bag of AN prills, the contents of which are later fed to the charging machine (or poured into a quarry hole).

For medium-scale operations, mechanical mixing in a concrete mixer lined with fibreglass, in small batches, can be quite effective. However, this procedure is not acceptable in Canada: a ribbon-type mixer or other specially-designed unit is required. Each type of equipment must be approved before use.

However, since AN/FO is now so freely available in pre-mixed form as a bagged product (see Fig. 33), at a marginally higher price, the practice of on-site do-it-yourself mixing of prills and oil is now considered disadvantageous.

For large scale work, bulk storage plants are usually established nearby, from which pre-mixed quantities of AN/FO may be bagged and despatched underground as required. For surface blasts, specialized bulk loading trucks draw from these storage plants and deliver to the blast site, either in pre-mixed form, or as separate ingredients (prills and oil) to be mixed and charged at the blast site. The latter of these bulk delivery vehicles are usually termed *mix trucks*. The mixture is charged into the blastholes either by an auger delivery system (see Fig. 1), or by a pneumatic unit. Where holes are moderately wet, the AN/FO may be loaded into a lay-flat polythene hole liner, or a special mobile de-watering system may be employed to pump water and sludge from the holes before charging.

Charging of Bulk Slurries

Mixing of AN slurries (SBA or SE) is a rather more exacting exercise; it is therefore usually handled by the suppliers. Their chief application is

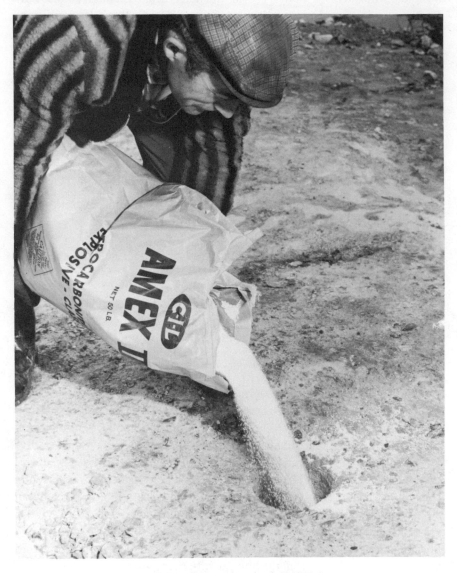

Fig. 33. Loading ready-mixed AN/FO into
hole by a simple pouring procedure.

Fig. 34. Pouring slurry into borehole by slitting the bag and shucking.

for surface blasts in wet holes or for breaking hard tenacious rocks (haematite or taconite).

Where slurries are used in small diameter short holes as for underground development headings and some stopes, soft plastic cartridged slurries are used.

But for long stope holes 5 inches in diameter and for large diameter holes in open cuts and quarries, bulk slurries are usually charged.

For small quarry operations, or in locations where bulk-mix-load services are not available, a series of **pourable slurries** is supplied in polyethylene packs that can be slit at the collar of the borehole and the contents shucked and poured down the hole (Fig. 34). Typical of these products are: DuPont "Pourvex Extra" and "Tovex Extra A", Gulf "Slurran 819", and CIL "Hydroflo".

For large scale operations, slurries are more advantageously placed by highly sophisticated *pump trucks,* usually operated by explosive manufacturers as a service to their customers. Many of these pump truck contractors provide for "programmed blasting", a system in which a loading formula can be preplanned for each hole or group of holes (with more than one mixture, if desired, for any particular hole), followed by fast economical accurate loading procedures. In this way, a single pump truck can load a number of mixtures into a hole at the one stop or visitation. Changes in the mixture formula can be effected by the operator, by using simple controls, during the pumping operation. By this means, slurry quantities and formulations can be tailored to fit the requirements of the hole pattern and the desired characteristics of the broken rock. Blasting performance is thereby optimized.

Some contractors also place the primers and fire the whole charge; thereby undertaking the whole of the blasting operation following the drilling of the holes.

Some pump trucks load with plant mixed slurries at the bulk plant depot; others have the capability of loading at the depot and mixing at the site (see Fig. 35).

Slurry explosives and blasting agents are denser than water; but it should not be assumed that pouring or gravity-loading into "down" holes will always effectively displace standing water and thereby provide a continuous column of slurry. To achieve this necessary objective, the holes should first be unwatered by pumping, and the slurry mix should then be pumped into the holes under pressure through a hose, even for vertical "down" holes of large diameter. The charging hose should first be lowered to the bottom of the hole and slowly withdrawn as the slurry progressively fills the hole.

Special unwatering pumps are now available.

Fig. 35. Pump truck for charging slurry holes as a programmed blasting service.

Priming of Blasting Agents and Slurry Explosives

The notoriously low initiation sensitivity of AN/FO (and slurries) predicates the need for great care in the priming of such charges.

For normal underground holes up to $2^1/_2$ inch diameter, a full cartridge of high velocity ammonia gelatine 60% , with a No. 6 detonator, is a sufficient primer in a well-mixed, well-placed, and well-stemmed charge of AN/FO.

For larger holes, as in quarries, priming needs extra care, especially if the hole is wet, or if decked charges are used. A small quantity of high-strength booster is preferable to a larger amount of lower strength. The

VOD of the booster must be greater than the inherent VOD of the charge (whether AN/FO or slurry), for efficient detonation.

The best position for the primer is at either end of the charge, i.e., with collar- or bottom-priming. Unless deckloaded, the practice of stringing additional boosters along detonating cord throughout the charge column is disadvantageous. It is important however to ensure good contact (without air space) between the primer and the charge. Nevertheless, the number and position of primers in a bench hole provide a continuing topic of research, experiment and debate.

Even with bottom-priming, the core loading of detonating cord should not exceed 25 gr/ft, in order to avoid premature low-order detonation of the charge laterally/radially.

With large diameter holes (say greater than 4 inches) the *shape* of boosters as well as the strength is important (Fig. 6). The diameter of such boosters should approach that of the borehole in order to provide that the major portion of the available energy is released axially (to propagate a strong detonation wave along the column) rather than radially. The use of Primacord as the sole detonant, even if it runs throughout the column, is not recommended, since it may cause deflagration rather than detonation of the charge.

Components of a Charge

Methods of formulating (and firing) a complete charge (for other than coal mines) are shown in Table 13.

Treatment of Misfires

A charge that has not exploded or has only partially exploded is classified as a "misfire" and is treated with the utmost caution.

The first precaution is to allow a safe interval of time (in some cases specified statutorily) between the report of a misfire and an approach to the face.

Before commencing and while removing the broken rock from the face, extreme care should be taken to identify unexploded cartridges or detonators, and to remove them to a safe place.

TABLE 13

Components of a Charge* (in other than coal mines)

Main Charge	Placement	Primer Details	Circuit Details	Method of Initiation	Method of firing	Method of sequential firing of several charges
Low explosive – Black powder in bulk	In borehole or cavity	-	-	Safety fuse	Light fuse with fuse igniter etc.	-
High explosive – NG-based (in cartridge form)	In boreholes, or coyote chambers,	Plain detonator incorporated in a cartridge (or a special booster)	-	Detonator crimped to safety fuse	Light fuse with fuse igniter etc.	Fuses docked to various lengths and ignited in order
Blasting agents and Slurry explosives (in bulk, prepacked cartridges, cans or bags)	or as blister or demolition charges		Igniter cord (IC)	Detonator crimped to safety fuse	Light IC with fuse igniter etc.	Fuses undocked, but connected to IC in order
		Electric detonator incorporated in a cartridge (or a special booster)	Series, parallel, parallel-series	Electric detonator connected to blasting circuit	Throw blasting switch on power lines, or operate exploder	Electric delay detonators
			Electric trunk line circuit	Electric detonators taped to DC downlines		
		Detonating cord laced through a HE cartridge (or a special booster)	Detonating cord trunk lines with connectors to DC "downlines"	Electric detonators taped to DC trunk line		M/S delay connectors
				Plain detonator (taped to DC) and safety fuse	Light fuse with fuse igniter etc.	

* The use of nonelectric Anodet, Primadet, Hercudet, and NONEL delay-systems are not included, because they are generally selfcontained systems. They are used only for initiating charges of blasting agents (AN/FO and AN slurries).

The face should be carefully examined for remaining "butts" of holes which may contain remnants of a charge. All such explosive remnants should be carefully removed before drilling commences. Holes should never be drilled into old butts.

In some states, a misfired hole can be rectified by drilling a hole near to but not closer than a specified distance from the misfire. The firing of this hole will either explode the misfired hole by sympathetic detonation or at least break it open and free the explosive.

Local regulations differ in all states, but they usually specify the procedure that should be carefully followed in each case.

Misfires can be caused by defective explosives (due mainly to deterioration in storage), damage to components while charging, water entering the detonator, discontinuity of the electrical circuit, insufficient current in the circuit, and by "cut-offs" (i.e., the explosion of a previous hole breaking out the collar of a hole and dislodging the primer).

AN/FO misfires are much less dangerous to resolve because AN is soluble and may be sluiced out of a butt with the water hose. However, if an NG primer is present, the misfired hole should be regarded as of the same degree of hazard as a conventional NG-based charge.

Tamping

The placing of cartridges of explosives in a small diameter hole, as in underground metal mining, is usually accomplished with a tamping stick (or rod). These are best made of wood, of about $7/8$ inch in diameter. Metal or metal-tipped tamping rods are dangerous. Drill steel must *never* be used. Rigid plastic tamping rods may also be hazardous because of the possibility of building up large charges of static electricity. Plastic materials used for tamping rods should therefore have suitable dielectric properties.

The operation of tamping a cartridged charge into a hole involves a certain amount of danger due to a premature explosion. Therefore great care should be exercised. Necessary precautions are as follows:

1. Holes should be thoroughly cleaned out before charging, to remove grit, sludge, duff, and obstructions that may cause undue impact pressure to be applied to get the cartridge into place.

2. Cartridge diameters should be less than the minimum diameter of the hole.

3. Cartridges should be firmly pressed into place and then enlarged by squeezing with the tamping rod to fill the borehole diameter. They should not be thumped or pounded.

4. Care should be taken not to dislodge the detonator, or damage the fuse or the leg wires.

5. Special care should be taken with the soft plastic cartridges of slurries. For charging holes in coal mines, see Chapter 20.

Stemming of Charge Holes

In order to localize the effects of the gases produced in the explosive reaction, the charge must necessarily be confined. A plaster charge to break large rocks or boulders is "mud-capped". Charges placed in boreholes must be sealed (or "stemmed") to prevent the gases from escaping before they make their contribution to the task of breaking the rock (see Glossary).

The most effective stemming material is sandy clay. It is usually pre-formed into cartridges and tamped into the hole above the charge.

For large quarry holes, drill cuttings are normally back-filled into the hole above the charge.

In coal mines, it is necessary to use inert incombustible material. Water stemming bags (Du Pont) have been used to seal the hole and to quench/cool the flame of the explosion, thereby reducing the risk of a coaldust/gas explosion.

An improved procedure is to use Trabant Gel-stemming ampoules.[6] These consist of a flame-resistant non-abrasive blow-moulded cylindrical sheath, filled and sealed (under factory conditions) with a stable inert gelatinous material containing water and organic substances. The ampoule not only effectively seals the hole, but also absorbs fumes and dust.

Fire-resistant polystyrene Tamping Plugs are also available, from Trojan.

Packing of Explosives

High explosives are usually bulk-packed in a fibreboard case of 50 lb weight or containing ten bags of 5 lb each. Thus in each bag, the number of wrapped cartridges depends on the diameter, length, and density of the particular explosive.

Cases are marked on each end with the name of the explosive, the diameter of cartridge, date, and batch symbol. Cases are lined with plastic to preserve the contents against deterioration during storage under normal conditions.

Rigid dynamite cartridges, although regularly $1^1/_8$ or $1^1/_4$ in × 8 inches long, are also supplied in $^5/_8$, $^7/_8$, $1^1/_2$, $1^3/_4$, 2, $2^1/_2$ and 3 in diameters in lengths of 8, 12, 16, 20 and 24 inches; and in even larger diameters for large quarry holes (Fig. 36). These explosives are supplied in prewaxed wrappers (to meet Class 1 fume requirements) in the smaller diameters for underground use. For surface conditions, sprayed paper cartridge wrappers are used, since these have better water resistance.

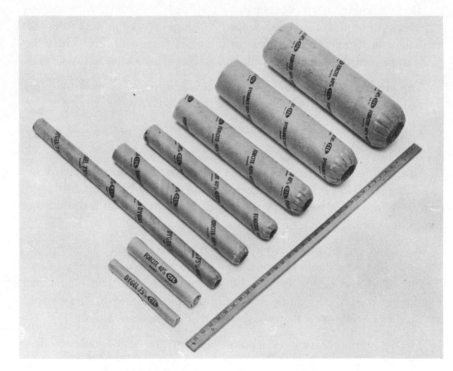

Fig. 36. Various sizes of rigid cartridged explosives.

For seismic prospecting with high explosives, where high heads of water are usually encountered, and for other underwater blasts, special packs have been devised. Threaded paper sleeves and rigid plastic shells in various lengths can be coupled together to form a rigid column, and thereby provide increased water protection, easier and improved positive charging techniques, and better propagation effects. For quarry blasting, large diameter cartridges are also supplied in a waxless tapered-crimp-ended (23G) heavy-walled spiral-wrapped tube (Fig. 3) that can be shipped bare without the need for a fibre-board case. These provide easier charging conditions and improved column propagation.

Cylindrical metal cans are used for blasting agents in coyote blasting, and for offshore seismic operations.

AN prills and AN/FO mixtures are usually supplied in bulk, or in 50 or 80 lb plastic, or plastic-lined multi-wall paper bags (Fig. 33). For large diameter quarry holes, AN/FO mixtures are also available in large diameter paper drums with a 23G tapered nose and lowering tapes (Fig. 3).

AN slurry mixtures are supplied in bulk by pump trucks or in plastic cartridges of $2^{1}/_{2}$ to 7 inches in diameter, which may be slit and loaded direct; or slit, shucked and loaded (where higher loading density and better coupling is desired. Fig. 34).

For regular seismic shots with blasting agents, a continuous rigid column of sealed threaded cans, provided with a nose cone (loading point), is formed and lowered into the hole. Individual cans contain 1 lb of material, packed 50 to the case (Fig. 2). A special primer can screws on to the top of column in the later stages of loading.

Storage of Explosives

Regulations governing the safe handling and storage of explosives are vested in the Chief Inspector of Explosives or his equivalent in each state or province (see Chapter 22).

Conditions governing the handling, storage, and use of explosives in mines, collieries, and quarries are laid down by the Chief Inspector of Mines.

As the regulations in each state or province are different, they cannot be quoted here because of space restrictions.

In addition to the need to store explosives securely to prevent their falling into the hands of unauthorized persons and to prevent their premature explosion, they should be stored in such a way as to prevent deterioration.

Explosives are perishable goods liable to deterioration after prolonged storage, especially under hot, humid conditions. They are packed and proofed to give maximum protection for long periods.

Fibreboard cases containing explosives are specially designed to give maximum strength (and therefore minimum distortion of cartridges) when stacked according to "This side up" instructions.

AN prills should be stored at a temperature not exceeding 90° F at all times. Repeated cycling through this temperature will cause the crystal lattice to become disrupted; particles thus become fractured and the desired physical characteristics of the prills are lost.

High explosives must be stored separately from detonators. Magazines must be well ventilated. Only non-ferrous tools should be used. Cases should be opened with wooden or copper wedges and mallets, as steel tools are not permitted.

Unwanted or deteriorated stocks should be disposed of in accordance with careful safe practice (see below).

The location of storage magazines in the United States should conform to the laws and regulations of the local state or municipal authority; and generally should follow the specifications set out in:

(a) Appendix A: Safe Location Specifications for Storage of Explosives (USA); otherwise cited and referred to as "American Table of Distances", or "American Table of Distances for Storage of Explosives.[1, 8, 9]

(b) Appendix B: Safe Location Specifications for Storage of Blasting Agent Materials (USA); otherwise cited as "Table of Recommended Separation Distances of Ammonium Nitrate and Blasting Agents from Explosives or Blasting Agents".[7, 8, 9, 10]

[This table lists recommended separation distances to prevent explosion of stores of ammonium nitrate and AN-based blasting agents (as the receptor) by propagation from nearby stores of high explosives or blasting agents (referred to as the donor)].

In Canada, when explosives are for the user's own operations, magazines for the storage of more than 150 lb of blasting explosives or more than 2000 detonators must be licensed by the Explosives Division of the Department of Energy, Mines and Resources. All magazines, regardless of capacity, containing explosives for sale must be licensed by that Department.[11] Magazine construction must comply with the minimum standards specified by the Department.[12]

Because cap-sensitive slurries are adversely affected by low temperatures, they should be stored at a minimum temperature of 40° F before loading into boreholes; or at least held in boreholes for a sufficiently long "residence time" before firing, to bring them to this temperature.[13]

Destruction of Blasting Materials

Explosives materials are perishable goods and may deteriorate after long periods of storage, even under approved conditions. Damaged or deteriorated explosives are dangerous and should be handled with extreme caution.[7, 14]

When it becomes necessary to destroy blasting materials, the advice of an explosives field representative, inspector of mines, or police officer should first be sought as to the safest method.

General Hints for Shotfirers or "Powder Monkeys"

This is the outstanding occupation that exemplifies the adage "Familiarity breeds contempt". Shotfirers should be well trained, should be steady, selfdisciplined types of men, and preferably non-smokers.

In the majority of mining operations, however, miners collect and look after their own stocks of explosives (for the shift in question), charge their holes, and fire their shots.

In other cases, however, it is possible and it is preferable to employ experienced and well-trained men who specialize in this class of work. This frequently happens in quarrying, tunnelling, shaftsinking, earthmoving, and colliery work.

The following hints apply both to experienced shotfirers and also to the regular miner who fires his own shots:

1. Avoid smoking or the use of naked lights.

2. Use care in handling.

3. Keep explosives dry, especially detonators and safety fuse.

4. Use safety fuse of adequate length.

5. Do not force or twist fuse into the detonator composition.

6. See that the fuse is freshly cut and cut squarely across.

7. Never crimp the detonator near the closed end. Use an approved crimper. Never use the teeth.

8. Do not bend, kink, or coil the safety fuse in tight circles.

9. When charging holes do not thump the cartridges with the tamping pole, but squeeze them firmly. Do not squeeze the primer but squeeze the stemming on to it.

10. Never use a metal or a metal-capped tamping pole.

11. When using electric primers, avoid kinking of the leg wires. Carefully free any accidental kinks. Do not damage the insulation.

12. When inserting an electric primer cartridge into the borehole, take care not to tug or jerk the wires; pull the wires *gently* to straighten them without kinking; avoid damage to wires when tamping.

13. When connecting up, bare and scrape the last 2 in of the ends of the leg wires; keep shunt in position as long as possible.

14. Carefully check all connections before running the connecting leads.

15. Never drag a firing cable along the ground—coil it carefully and carry it, paying it out as you go.

16. Always maintain personal possession of exploder key or key to mains firing boxes. Make sure the latter are locked (with fuses drawn and switch in the safe position) before approaching the face for charging.

17. Never return to a face after a blast unless:

    (1) You have the exploder key in your possession.

    (2) Mains firing box is locked in the safe position.

    (3) Leads are disconnected and short-circuited.

    (4) An adequate time interval has passed (usually 30 min for fuse firing).

18. Make sure all personnel have been withdrawn to a safe place before firing.

19. Be especially careful of misfires.

20. Report any misfires to the oncoming shift.

21. Never leave unused explosives lying about. Return them to the magazine as soon as possible.

22. Never charge a bulled hole until it has cooled down.

23. Never bull a hole next to one already charged with explosive.

24. In general, observe faithfully the local regulations laid down for the safe use of explosives, especially those relating to misfires.[15]

Causes of Explosives Accidents

Positive care should always be exercised in the use of explosives, as accidents associated with them are dramatically horrible. The following is a list of common causes of explosives accidents; each cause infers some lack of care or disregard of regulations or established safe practice.

1. Delaying too long in lighting fuse.

2. Drilling into explosives.

3. Premature firing of electric blasts.
4. Returning too soon after blasting.
5. Inadequate guarding.

6. Unsafe practice during transport, handling and storage.[9]

7. Improper handling of misfires (see above).

8. Using fuse too short in length.

9. Improper tamping procedure.

10. Smoking during handling of explosives.

Basic information regarding accidents that might result from electric currents (such as lightning strikes, electric storms, static electricity, radio frequency currents, and other stray currents) is also published.[16, 17]

References

1. E. W. S. COLVER, *High Explosives*, 2nd. ed. (London, Technical Press, 1938).

2. W. O. SNELLING and W. C. COPE, "The Rate of Burning of Fuse, as Influenced by Temperature and Pressure", *U.S. Bureau of Mines, T. P. 6*, 1912.

3. "Pneumatic Loading of Nitro-carbo-nitrates; Static Electricity, Fumes and Safety Handling". Technical Report (undated), Atlas Chemical Industries Inc.

4. E. L. LITCHFIELD, M. H. HAY and J. S. MONROE, "Electrification of Ammonium Nitrate in Pneumatic Loading". *U.S. Bureau of Mines, R. I. 7139*, 1968.

5. R. A. DICK, "Puzzled about Primers for Large-diameter ANFO Charges? Here's some Help to End the Mystery", *Coal Age*, August 1976.

6. "CPMS Trabant Gel Stemming", Contractors' Plant and Material Supply Co. Ltd., Llanelli, U. K.

7. "Safety in the Transportation, Storage, Handling and Use of Explosives", Institute of Makers of Explosives, Pub. No. 17, 1977.

8. "The American Table of Distances", Institute of Makers of Explosives, Pub. No. 2, 1977.

9. "Manufacture, Storage, Transportation, and Use of Explosives and Blasting Agents, 1970", National Fire Protection Association, Booklet No. 495, 1973.

10. R. W. VAN DOLAH, "Large-scale Investigations of Sympathetic Detonation", *Ann. N. Y. Acad. Sci.*, Vol. 152, Art. 1, October 1968, pp. 792—801.

11. "Storage of Explosives", Dept. of Energy, Mines and Resources, Explosives Division, Ottawa, 1972.

12. "Standards for Blasting-Explosives Magazines", Dept. of Energy, Mines and Resources, Explosives Division, Ottawa, 1971.

13. "Heated Explosives Magazines", Canadian Industries Ltd., 1975.

14. "How to Destroy Explosives", Institute of Makers of Explosives, Pub. No. 21, 1970.

15. "Instructions and Warnings", Institute of Makers of Explosives, Pub. No. 4, 1973.

16. "Radio Frequency Energy", Institute of Makers of Explosives, Pub. No. 20.

17. "Safety Guide for the Prevention of Radio Frequency Radiation Hazards", Institute of Makers of Explosives, Pub. No. 20, 1977.

18. American National Standards Institute, Standard C 95.2, 1966.

PART II

BREAKING GROUND
WITH EXPLOSIVES

ABBREVIATIONS

AWG	American Wire Gauge
FGAN	Fertilizer grade ammonium nitrate
ft	foot, feet
ft^3	cubic foot (feet)
ft^3/lb	cubic feet per pound
ft/sec	feet per second
ft^3/ton	cubic feet per ton
gm	gram(s)
gm/cc	grams per cubic centimetre
gr/ft	grains per foot
in	inch(es)
lb	pound(s)
lb/ft^3	pounds per cubic foot
lb/in^2	pounds per square inch
lb/ton	pounds per ton
min	minute(s)
mm	millimetre(s)
msec	millisecond(s)
oz	ounce(s)
oz/ft^2	ounces per square foot
oz/yd^3	ounces per cubic yard
sec/ft	seconds per foot
USSS	United States Standard Sieve

CHAPTER 12

Introduction

General

Military explosives are not dealt with in this book.

Explosives are used in the industrial sphere chiefly for fracturing ores and rocks either in the recovery of economic minerals or for the making of excavations within a rock mass. The former includes the winning of road metal and building materials by quarrying, and of ores and minerals by open cut and underground stoping operations, where the rock or ore in place is too hard and tough to loosen by mechanical or presently developed hydraulic methods. The latter refers to the excavation of road cuttings, tunnels, city building excavations and submarine passages in civil engineering practice and of underground development openings (such as shafts, drifts, crosscuts, raises, and winzes) in mines.

Industrial explosives are also used (but to a minor extent) for demolishing foundations and concrete or masonry structures, for breaking heavy castings, and for general agricultural usage (see Chapters 17 and 21).

Industrial explosives are therefore vital to modern civilization and have contributed greatly to our current standard of living.

Their main use is in breaking out selected sections of hard rock and causing the "fragmentation" of such rock in the process. There is as yet no other way of doing this work practically and economically.

In order so to fracture rock (or to "break ground"), the explosive charge must be placed within the rock and at a suitable distance behind the rock face. For this purpose, openings must be made *into* the rock either by drilling holes (with or without subsequent "bulling" or chambering) or by excavating chambers (as for very large charges in coyote blasting).

The rock mass must also have one (or more) free faces, i.e., it must be exposed or open on one or more planes more or less at right angles to that from which the drilling is done. This provides a rock/air interface termed a "free face". The rock is blasted in the direction of the free face.

This is necessary because (a) broken rock occupies a much greater space than when it is in the solid mass, and it must therefore have room to move or expand (see Fig. 37), and (b) it thus allows the rock to be broken in tension (more effectively) rather than in shear (under "tight" conditions with no free face).

When a free face does not exist (as in the case of tunnel or mine development headings), it must be provided in one of the following ways (Chapter 15):

1. By explosives in cut holes (centre, wedge, or draw cuts).

2. By explosives in cut holes in conjunction with free (uncharged) holes (as in burn cuts).

3. By drilling a large diameter hole or slot (an extension of the burn cut principle).[1]

4. By excavating a large diameter hole with a shaped charge.

5. By the use of high pressure pulsed water jets (still under development); or

6. By mechanically cutting a kerf (as in a coal face).

In Figure 37, the distance DC (or FD) is known as the "burden distance" (see Glossary). It represents the line (or path) of least resistance, along which the charge can more easily break the rock; or along which the burden of rock is least. The free face BC is actually a rock/air interface. Some writers regard the face from which the drilling is done as one free face, in which case, *two* (or more) free faces are required.

References

1. R. GUSTAFSSON, *Swedish Blasting Technique* (Gothenburg, 1973), pp. 131—136.

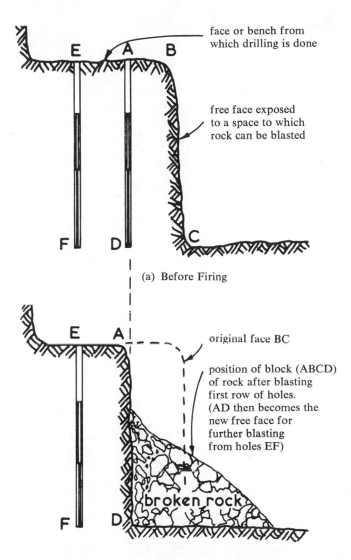

face or bench from
which drilling is done

free face exposed
to a space to which
rock can be blasted

(a) Before Firing

original face BC

position of block (ABCD)
of rock after blasting
first row of holes.
(AD then becomes the
new free face for
further blasting
from holes EF)

broken rock

(b) After Firing one Row of Holes

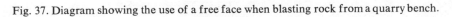

Fig. 37. Diagram showing the use of a free face when blasting rock from a quarry bench.

CHAPTER 13

Blasting Theory

Factors Involved

Explosives are used for breaking ground only when cheaper effective means are not possible or available.

In breaking ground, the following conditions have an effect on blasting operations:

1. Presence and extent of one or more free faces.

2. Tenacity or cohesive strength of the rock (i. e., its resistance to blasting).

3. Structure of rock (massive, stratified, jointed, or fissured).

4. Type, strength, and nature of the explosives used.

5. Method of initiation.

6. Type of detonation (instantaneous or sequential — regular delay or short-delay).

7. Whether broken rock falls or is to be lifted.

8. Size, type, and depth of shothole or chamber.

9. Depth to which hole is loaded.

10. Proportion of burden distance to length of hole.

11. Loading density and degree of confinement.

12. Subsequent method of removal of spoil.

13. Degree of fragmentation required.

14. Coupling ratio effected.

15. Temperature of explosive charge (if a slurry).

16. Proximity or otherwise to important installations or built-up areas.

17. Amount of air concussion allowable.

18. Quantity of rock to be broken.

19. Whether the holes are in water.

20. Ultimate shape of excavation.

21. Ultimate use or purpose of excavated material.

22. Climatic conditions.

23. Safety of human life.

24. Statutory regulations.

ATCHISON[1] epitomizes these rock fragmentation factors rather more succinctly as:

(1) **Explosive parameters,** such as density and detonation velocity (detonation pressure), detonation impedance, gas volume and available energy.

(2) **Charge loading parameters,** including charge diameter and length, stemming, coupling, and type and point of initiation; some of these may directly influence the explosive parameters.

(3) **Rock parameters,** particularly density, propagation velocity, characteristic impedance, energy absorption, strength (compressive and tensile), texture and lack of homogeneity or isotropy.

Blasthole Diameter

The choice of borehole diameter in quarry blasting is critical since it affects the drill equipment specifications, burden distance, spacing distance, explosives distribution, and generally the efficiency and economics of the complete operation.[2] For this reason, the choice of borehole diameter should not be restricted to considerations of minimum drill/blast costs in isolation. Other parameters can be varied once the optimum blasthole diameter has been determined.

The most important considerations in determining optimum borehole diameter for a given operation are:

(a) Rock parameters (see above).

(b) Explosive parameters (see above) that are related to charge diameter.

(c) Local restrictions connected with the proximity of built-up areas.

(d) Production factors, related to size of loading, haulage, crushing, screening equipment and scale of production.

It is important to match drill equipment specifications (and therefore blasthole diameter) to the later categories of production equipment, since the cost of purchasing, operating and maintaining the latter may be of paramount importance, depending mainly upon (a) the tenacity and abrasiveness, and (b) the ultimate end use of the material to be broken.

In theory, it should cost more per inch of diameter to drill larger diameter holes, but recent installations have demonstrated that the massive proportions of metal and power in the larger drill rigs achieve penetration rates comparable with those of smaller diameter capacity. There has therefore been a recent pronounced trend towards larger diameter blastholes in order to achieve the economies associated with improved labor productivity, extended blasting patterns (see Glossary), and lower density low-cost blasting agents. However, when toe problems increase and higher costs of secondary blasting and load/haul/crush operations due to poor fragmentation ensue, then the economic limit of large diameter blastholes has certainly been reached or exceeded.

Such limits depend upon (a) the comparative ease with which the particular rock formation will become fragmented, and (b) the economic degree of fragmentation required, consistent with the end use of the material to be broken.

The determination of an optimum blasthole diameter should involve adequate consideration of these factors.

Choice and Quantity of Explosive

This mainly depends on the tenacity and other physical characteristics of the rock being broken. The presence of bedding planes, joint planes, and planes of weakness also influences the choice.

Further, the diameter of the borehole has an important bearing on the question. Large diameter boreholes or small diameter holes that are "chambered" would call for the use of lower strength low-density explosives.

Small diameter boreholes on the other hand, especially in the harder more tenacious formations, call for high density explosives.

The quantity used is a function of the hole diameter, hole depth, burden distance and spacing distance for any particular advance, after the type, strength, and density of explosive has been determined for the particular conditions.

In determining the type, strength, and quantity of explosive to be used for a given purpose, the over-riding considerations are:

1. Applicability to the particular job.

2. Safety of life and property.

3. Efficiency (the usual yardstick is the "powder factor" expressed in lb per ton of ore or rock broken to the required size. This ratio varies generally from 0.2 to 1 lb per ton, depending principally on the tenacity of the rock).

4. Economy (based upon the total costs of the operation, including drilling, firing, loading, haulage, crushing, screening and product storage).

Fragmentation

Poor fragmentation usually results in high secondary blasting costs and high costs of loading, haulage, primary crushing and general maintenance.

Fragmentation can be improved in several ways:

1. Shallower holes or better distribution of the explosive charge over the length of the hole.

2. Shorter spacing distance between holes.

3. Shorter burden distance.

4. Use of an explosive that gives less brisance and greater gas production.

5. Use of short-delay detonation.

Analysis of Theories

Early formulae for the calculation of blasting charges were evolved for military blasting operations. In these formulae, the burden distance alone was taken as the variable factor while a correction factor embraced the effects of type of explosive, blasting resistance of the rock, depth and diameter of drill holes, and charging depth. This meant that the correction factor was useful only for a limited range of conditions. The best of these formulae was due to Belidor.[3] Bendel[4] has published a critical discussion on the efficacy of these blasting formulae.

The first atomic bomb blast provided an impetus to research on blasting problems, giving rise to the development of the Reflection Theory[5] (see Chapter 2).

A review of other treatments of blasting design is included here for the information of students.

1. O. ANDERSEN[6] shows that the main considerations in breaking ground are (a) the burden distance and (b) the strength, quantity, and distribution of the explosive charge. The first is a function of the square root of the projected area of the blast hole, and is *independent of the nature of the rock;* the second takes care of the physical characteristics of the particular rock. Both are to some extent modified (when more than one hole is fired) by the spacing distance between holes and the degree of fragmentation required.

The following simple empirical formula for a single hole gives the burden distance in terms of the projected area of the hole, viz:

$$B = c \sqrt{DL}$$

where B = burden distance, ft

D = hole diameter, ft

L = length of hole, ft

c = constant determined empirically.

It was also shown that sufficiently accurate values were given for most practical purposes by quoting the hole diameter in inches and by taking $c = 1$.

Thus the formula was applied in practice as

$$B = \sqrt{dL}$$

where d = hole diameter, in inches

Over a range of blast holes from 1 inch diameter, 4 ft deep to 9 inches diameter, 100 ft deep, burden distances have been calculated, and are shown in Table 14.

TABLE 14
Burden Distances Calculated by the ANDERSEN Formula

Diameter of Hole (in)	Length of Hole (ft)								
	4	9	16	25	36	49	64	81	100
1	2	3	4	5	6	7	8	9	10
2	3	5	6	7	9	10	12	13	14
3	–	–	7	9	10	12	14	16	17
4	–	–	8	10	12	14	16	18	20
5	–	–	9	11	13	16	18	20	22
6	–	–	–	12	15	17	20	22	25
7	–	–	–	13	16	19	21	24	27
8	–	–	–	14	17	20	22	25	28
9	–	–	–	15	18	21	24	27	30

This formula, unlike the others set out in this chapter, has the particular advantage that the burden distance can be determined from readily available physical dimensions, without the need for the separate trial assessment of other factors.

Burden distances calculated from the above formulae have been shown to agree very closely with those employed in actual practice; and especially for the wide range of commonly occurring rock types ranging between plastic clay shales and mudstones and incompetent overburden on the one hand, and dense tough hard tenacious haematite and taconite on the other. Within the above limits of rock types it is highly recommended for both undergraduate use, and indeed for most practical applications.

But when aluminized AN/FO and slurries are used, in well-stemmed holes with adequate primage and excellent coupling, the explosive parameters are enhanced. The increase in density, VOD, and heat of reaction produces a much higher detonation pressure. As a result, an increased value of c will usually be applicable, yielding a greater practicable burden and spacing distance (extended blasthole geometry). But field trials will normally be necessary to determine the extent of the increase possible, consistent with fragmentation requirements.

When more than one hole is fired at a time, these burden distances can influence the amount of spacing adopted between adjacent holes in any row. Modified formulae covering this effect have been adduced by ANDERSEN.

2. G. E. PEARSE[7] offers a burden formula based on the physical characteristics of rock and type of explosive:

where $$B = Kd \sqrt{P_s/T}$$

   B is maximum burden, in inches

   K is constant depending on rock characteristics (0.7 to 1.0);

   d is hole diameter, in inches

   P_s is reaction stability pressure of explosive, lb/in^2.

   T is ultimate tensile strength of rock, lb/in^2

The reaction stability (peak explosion) pressure of particular explosives has been calculated by TAYLOR.[8]

3. R. L. ASH[9, 10, 11] has propounded a simplified modification of the Pearse formula, in which he replaced the factors K and $\sqrt{P_s/T}$ by a burden ratio, K_b; from empirical field tests, the factor K_b ranges from a value of 20 to 40, averaging 30.

He bases his observations on the dependence of burden distance on a combination of rock and explosives parameters. The mean value of $K_b = 30$ corresponds by and large to blasting conditions in which the rock density averages 2.7, and the explosive has a density of 1.3 and an average VOD of 12,000 ft per second. Actually, the K_b values range from about 25 for AN/FO to 35 or more for dense slurries and gelatines, under average rock conditions. Various formulae have been adduced by ASH to adjust K_b values for more specific conditions, such as variations in rock types, in density of explosive, and for the degree of fragmentation and rock displacement desired.

It will therefore be seen that the ASH formula caters for refinements not available in the ANDERSEN formula. But it necessarily involves a number of separate determinations to arrive at a suitable K_b value. These may be assessed from convenient tables.

As a recommended practice, the ANDERSEN formula can be used as a first step. Where further refinements in blasting design are considered necessary, the next step is to evaluate the parameters, calculate the K_b value, and apply the ASH formula.

The maximum burden distance is given by

$$B = K_b \times d/12 \text{ (ft)},$$

where d is hole diameter, in inches.

From this simplified but very useful formula, ASH has also derived relationships for assessing the spacing distance (S), the length of hole (H), the length of subgrade drilling (J), and the collar (stemming) distance (T), all in terms of the burden distance (B). These are set out below.

Spacing distance, $S = K_s B$

Hole length, $H = K_h B$

Subgrade drilling depth, $J = K_j B$

Collar distance, $T = K_t B$

where K_s, K_h, K_j and K_t are the spacing ratio, hole length ratio, subgrade drilling ratio, and collar distance ratio respectively. All distances are expressed in feet, with ranges of $K_s = 1$ to 2; K_h from 1.5 to 4; K_j generally from 0.2 to 0.4; and K_t from 0.7 to 1.

All these factors are useful in calculating a complete blasting design programme.

4. FRAENKEL[12] has derived a somewhat similar formula to that of ANDERSEN[6] in establishing a rock characteristic which he calls "resistance to blasting".

$$B = \frac{RL^{0.3} \times l^{0.3} \times d^{0.8}}{50}$$

where

B is burden distance, in metres

R is resistance to blasting (an experimental factor varying from one to six for all rock types)

L is length of hole, in metres

l is length of charge, in metres

d is diameter of hole, in mm.

This formula has been derived from the use of Swedish 35 per cent LFB dynamite. In practice, the following approximations are used:

B is reduced to $0.8B < 0.67L$,

l is taken as $0.75L$,

S (the spacing distance) is not greater than $1.5B$.

5. LANGEFORS has produced a comprehensive series of formulas based largely upon the work of FRAENKEL.[13] He quotes the maximum theoretical burden distance as

$$B_{max} = (45 \times d) \quad \text{(metres)}$$

where d is the hole diameter in mm.

Then he proceeds to develop a measure of the "practical burden distance" by modifying B_{max} for variables that might be expected to occur in practice, such as faulty drilling (errors in hole spotting and in hole deviation). Included in the calculations are associated determinations of other factors such as the length of blast hole below grade, the total hole depth, bench height, hole spacing, height and weight of bottom charge, height and weight of column charge, uncharged height, specific charge, specific drilling length, and width of the face.

GUSTAFSSON offers a series of examples and tables listing these various calculations under prescribed rock conditions.[14] However, it is difficult to represent the widely varying rock characteristics in any mathematical model when the rock is not homogeneous. Rock is never really homogeneous.

6. **Crater Theory** (PEELE).[15] If a vertical hole is placed normally to a horizontal surface of earth or rock and charged with explosive, it may blow out a conical shaped crater, the sides of which meet the horizontal surface (or free face) at $45°$.

In such a case, the line of least resistance is given by L, the depth of the hole. The volume of the crater:

$$V = 0.33L \times \pi L^2 = L^3 \text{ (approx.)}$$

In practice, the volume of rock loosened is taken as:

$$V = mL^3$$

where $m = 0.4$ for tough, soft rock and 0.9 for hard brittle rock.

Crater theories are generally based on a point source of explosive. This means that although they are useful in laboratory studies, it is not practicable to reproduce the effects so observed under actual field conditions. Nevertheless, the principle in basic. The 90° conical shape was first propounded by VAUBAN in 1704.[16]

7.**The Livingston Crater Theory**[17] has had wide acceptance.

LIVINGSTON developed his "strain energy equation" for crater charges as set out below:

$$N = EW^{1/3}$$

where N is the critical depth, in ft, of a charge

W is weight, in lb, that would just cause the rock surface to fail, and

E is the strain energy factor derived empirically.

He then modified this equation by reducing the charge depth to give good fragmentation by expressing it in the following form:

$$d_0 = \Delta EW^{1/3} \text{ (ft)}$$

where d_0 is the optimum depth (or indicated maximum burden distance)

Δ is the optimum depth ratio (d_0/N)

W is the weight of charge, lb.

Useful values of d_0 depend upon the value of E which varies for different rock characteristics, as discussed by BAUER.[18]

GRANT has established the Dow Crater Formula by adopting a standardized cylindrical volume (6 inch diameter × 36 inch) of explosive charge, instead of the weight W used by LIVINGSTON.[19]

This simplifies comparative field performance testing of explosives.

8. **The Thermohydrodynamic Theory:**[20] The mathematical and physicochemical analysis associated with the exposition of this theory is too

involved for inclusion in an elementary monograph. It is useful only as a research tool; or to determine the theoretical energy output of particular explosives.

Quarry Blast Design

Because of the complex problems involved, none of the above theories or formulae can be expected to give precise values of burden distance; nor is this essential under practical conditions of blasting. They do, however, provide an opportunity to set up initial blasting patterns; or to check existing practices where these are suspect; or where reduced blasting costs and better blast results are sought. A simplified analysis of the main design features in quarry blasting is given below.

1. Determination of burden distance

The maximum amount of work in breaking out a rock face has to be done at the bottom of the borehole where the charge must supply sufficient energy to shear the rock at floor level in addition to breaking it along the line of holes and displacing the mass of rock. The **burden distance** is the distance between the bottom of the hole and the free face, and this corresponds with the line of least resistance.

Taking an extreme case, where the burden distance exceeds the depth of the hole, the result of firing a suitable charge at X will be a conical-shaped hole as in Figure 38. Where the burden distance equals the depth of the

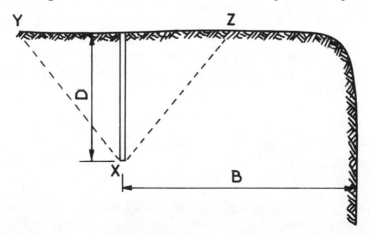

Fig. 38. Probable blasting effect where burden distance exceeds hole depth.

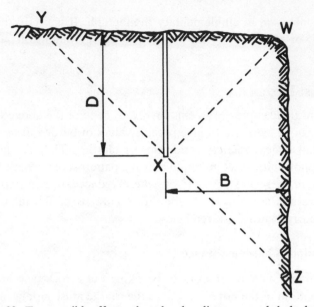

Fig. 39. Two possible effects where burden distance equals hole depth.

hole, the result may be as shown in Figure 39. In this case, the break would be along the lines YXZ, if the charge was strong enough; however, in practice, it would tend to blow a crater YXW.

Figure 40 shows how the rock will break if the burden distance is less than the hole depth, and the hole is column-loaded but does not reach the lower quarry bench level. If the hole is loaded up to point Xn, then the section of rock broken corresponds to the parallelogram XXnYnZ. However, for practical purposes, it is necessary to preserve the continuance of a horizontal floor or bench level (to accommodate a roadway), and therefore XZ should be horizontal as in Figure 41.

Now Figure 42 shows the corresponding stress pattern for a concentrated charge at X. Only a small proportion of the stress wave is reflected in tension, and therefore the rock must be broken along XZ mainly in shear (in which property it is about ten times as strong).

Hence, an inordinately larger quantity of explosive is required at X to break effectively along XZ. If it fails to do so, a "toe" will be left, as in Figure 43. In this event, before further (economic) blasting operations can proceed, the face and bench must be "squared up". This can only be done by *ad hoc* drilling and blasting, involving a disruption to the operational plan, and a general increase in costs. This sort of problem occurs frequently in many quarries where the rock has a high shear strength.

The persistence (or regular occurrence) of a toe in quarry blasting signifies that the burden distance is too great; or that the bottom charge (the explosive in the bottom of the hole) is not strong enough; or that the hole does not extend far enough below bench level to accommodate a sufficient explosive charge to shear the rock cleanly.

In order to overcome this problem, KOCHANOWSKY[21, 22, 23] has shown that, in theory, the holes should be drilled at an angle of 45°, so that the maximum effect of breakage in tension can be derived (see Fig. 44).

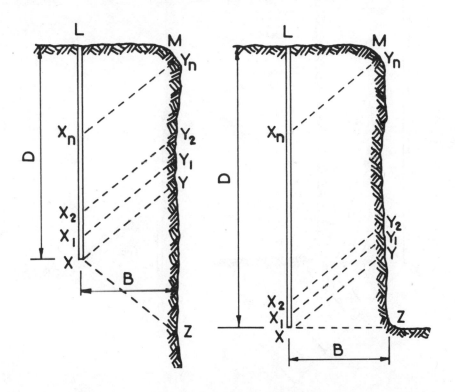

Fig. 40. Blasting effect where hole is column-loaded, and bottom of hole does not reach the lower

Fig. 41. Blasting effect where hole is column-loaded, and bottom of hole reaches the lower quarry bench.

Such a practice has many useful advantages; it is now gaining ground (except that, for greater ease of drilling and loading, the inclination of the holes is restricted to 60°—65° as a compromise).

KOCHANOWSKY[23] lists the advantages claimed for angle drilling as follows:

(a) Safer operation for both men and equipment.

(b) Improved fragmentation, and greatly reduced secondary blasting cost.

(c) Negligible toe problem.

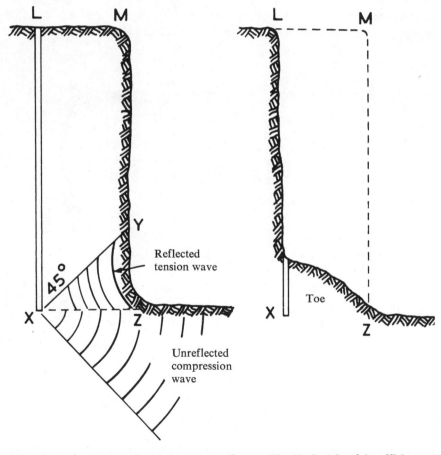

Fig. 42. Stress pattern for a concentrated charge in bottom of hole.

Fig. 43. Result of insufficient explosive energy release at bottom

(d) Backbreak difficulties practically eliminated.

(e) Progressive reduction in explosive consumption with increased inclination from the vertical.

(f) Lower drilling footage per ton of rock broken.

(g) Less vibration.

(h) Significant reduction in drilling, blasting, loading, and crushing costs.

From Figure 41 it will be seen that:

(i) The chances of failure to break the "toe" vary directly with the burden distance.

(ii) No explosive should be placed nearer to the collar of the hole than the distance to the free face XZ; otherwise, in most cases it will be wasted, or it may cause rock to "fly", or cause unnecessary backbreak.

The toe burden that can be broken satisfactorily in a deep hole blast depends, among other things, upon the weight of the charge that can be accommodated in the bottom portion of the hole. Actually, the weight of charge in a borehole is proportional to the square of the diameter.

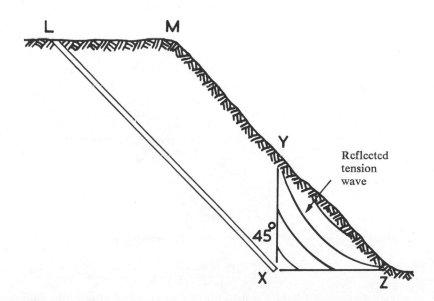

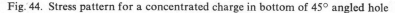

Fig. 44. Stress pattern for a concentrated charge in bottom of 45° angled hole

2. Determination of Spacing Distance

Figure 45 is a plan showing the wedge of rock which should be broken by a charge at F; the volume of rock broken to the free face EG by a vertical hole at F is the square of the burden multiplied by the depth. An increase in the burden of 50 per cent calls for an increase in the charge of 125 per cent, to cater for a similar increase in volume of rock broken.

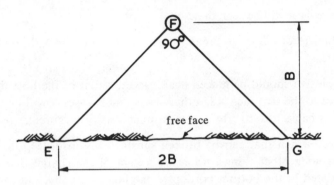

Fig. 45. Plan diagram of volume of rock broken to the free face by a single vertical hole at F.

The jointing of the rock is the most important factor in determining the spacing between holes in a row. If the hole spacing is equal to twice the burden the result should be as in Figure 46. The triangle of rock which lies between the bursting angles of the two charges at M and N would, in theory, remain unbroken.

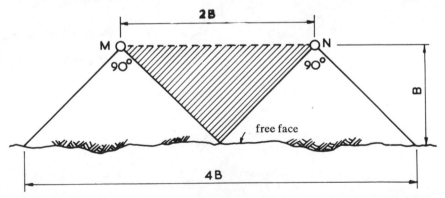

Fig. 46. Plan diagram of breakage pattern for two holes spaced at twice the burden distance.

In practice, however, the rock would be partly dislodged or poorly fragmented. It would, nevertheless, depend on the conditions existing at a particular site.

But if the spacing is equal to the burden, as in Figure 47, the bursting
angles would intersect at P, leaving the rock in the triangle MNP outside
the theoretical bursting angles. In practice, however, the area broken will
be LMNO. Since explosives exert force in all directions, this triangle should
be well fragmented and the face from M to N broken cleanly. Where blast
holes are spaced at intervals which are too wide, the face is frequently left in
a ragged condition, and, in massive rocks, will often present a "scalloped"
appearance. In addition, the rock is "fragmented" to a lesser degree.

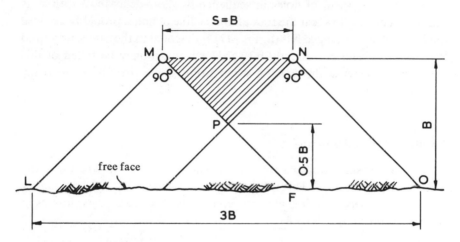

Fig. 47. Plan diagram of breakage pattern for two holes spaced at the burden distance.

Experience tends to show that changes in the burden distance have a
much more pronounced effect on the fragmentation, on the tendency to
leave an unbroken "toe", and on general quality of breakage than changes
in spacing distance. Therefore, if an increase in the tonnage per foot of hole
is desired for economic reasons, it is better to extend the spacing between the
holes rather than increase the burden.

In all references to the geometry of blasting patterns it is customary to
express the burden distance and spacing (in feet) as "B by S" or "B x S",
always quoting the burden distance first. Normally, the value of S varies
from B to 2 B, averaging 1.5 times B.

3. Methods of firing

The principle of firing a number of charges simultaneously so that they will mutually assist one another is important in practically all surface blasting operations. But simultaneous firing of charges has placed a restriction on the size of the blast that can be fired in many locations, since the vibration produced often results in complaints and, in some cases, damage to adjacent property and installations.

By using short-delay firing methods, the principle of mutual assistance is retained, and without increasing the vibration the size of the blast may be increased significantly.

When firing a row of holes in sequence by short-delay means, there is a greater tendency to shear the rock along the line of holes, probably because the line of least resistance NP (in Fig. 47) is diagonal to the previously fired hole, as at M, and not directly forward. Advantage may be taken of this tendency to increase the spacing when firing single rows by short-delay methods.

4. Calculation of the charge

Firstly, for successful blasting the borehole must be drilled to give the correct burden distance. This is most important and should not be a matter of guesswork. Secondly, it is advantageous to use high density explosives in the bottom portion of the hole.

A simple formula for calculating the amount of explosive required for a single hole is $C = B \times S \times D/60$, where C is the charge weight, in lb, B is the burden distance, in ft, S is the spacing distance, in ft, and D is the hole depth, in ft. This formula gives a powder factor of 0.225 lb/ton, for rock of 13.5 ft³/ton (see Appendix D).

It is generally applicable for high density explosives (dynamites and slurries).

With large diameter holes (5—12 inches), where high explosives are used, the charge would usually be deck-loaded, in which case two-thirds of the total charge should be placed in the bottom of the hole. The charge may be distributed or placed to suit local conditions. Deckloading may not necessarily be required when other explosives of a lower loading density (e. g. AN/FO) are employed.

A useful report to assist quarry operators has been prepared by PUGLIESE.[11]

A noteworthy approach to the optimization of burden/spacing patterns has been propounded by LANG and FAVREAU.[24] They have developed a computer model in which all relevant parameters of rock properties, explosive energy outputs and blast geometry factors are matched in terms of rock movement time and cost elements to yield optimum effects quantified by numerical values, within the natural limitations of rock structural variation in the immediate blast area.

References

1. T. C. ATCHISON, "Fragmentation Principles", Chapter 7.2 of "Surface Mining", ed. E. P. Pfleider. Seeley W. Mudd Series (New York, AIME, 1968).

2. R. A. DICK and J. J. OLSON, "Choosing the Proper Borehole Size for Bench Blasting", *University of Minnesota, 31st Annual Mining Symposium*, 1970.

3. B. F. BELIDOR, *Nouveau Cours Mathématique à l'Usage d'Artillerie et du Génie (Paris:* 1725), p. 505.

4. L. BENDEL, *Das Sprengen im Fels* (Lucerne, Haas Verlag, 1942).

5. K. HINO, "Fragmentation of Rock Through Blasting", *Jour. Ind. Explosives Soc. Japan*, Vol. 17, No. 1 (1956), 1—11.

6. O. ANDERSEN, "Blast Hole Burden Design", *Proc. Aus. I. M. M.*, No. 166—7 (1952).

7. G. E. PEARSE, "Rock Blasting", *Mine & Quarry Engineering*, January 1955.

8. J. TAYLOR, *Detonation in Condensed Explosives* (London: O. U. P., 1952).

9. R. L. ASH, "The Design of Blasting Rounds", Chapter 7.3 of "Surface Mining", ed. E. P. Pfleider, Seeley W. Mudd Series (New York, AIME, 1968).

10. R. L. ASH, "The Mechanics of Rock Breakage", *Pit and Quarry*, Vol. 56, Nos. 2, 3, 4, 5.

11. J. M. PUGLIESE, "Designing Blast Patterns Using Empirical Formulas", *U. S. Bureau of Mines, I. C. 8550*, 1972.

12. K. H. FRAENKEL, "Factors Influencing Blasting Results", *Manual on Rock Blasting* (Stockholm, Aktiebolaget Atlas Diesel, 1952), Vol. I, Art. 6, No. 2, p. 15.

13. U. LANGEFORS and B. KIHLSTROM, *The Modern Technique of Rock Blasting* (New York, Wiley, 1964).

14. R. GUSTAFSSON, *Swedish Blasting Technique* (Gothenburg, 1973).

15. R. PEELE, *Mining Engineers' Handbook* (3rd ed.), pp. 5—11.

16. S. S. SALUJA, "Mechanism of Rock Failure under the Action of Explosives", *Ninth Symposium on Rock Mechanics*, Golden, Colo. April 1967.

17. C. W. LIVINGSTON, "Fundamental Concepts of Rock Failure", *Colo. School of Mines Quarterly*, Vol. 51, No. 3.

18. A. BAUER, "Application of the Livingston Theory", *Quarterly of the Colorado School of Mines*, Vol. 56, No. 1 (January 1961).

19. C. H. GRANT, "Simplified Explanation of Crater Method", *Engineering and Mining Journal*, Nov. 1964.

20. M. A. COOK, *The Science of High Explosives* (New York, Reinhold, 1958).

21. B. J. KOCHANOWSKY, "Blasting Research Leads to New Theories and Reductions in Blasting Costs", *Mining Engineering*, September 1955.

22. B. J. KOCHANOWSKY, "Inclined Drilling and Blasting", *Mining Congress Journal*, November 1961.

23. B. J. KOCHANOWSKY, "New Developments in Drilling and Blasting Techniques", *Engineering and Mining Journal*, Vol. 165,, No. 12 (December 1964).

24. L. C. LANG and R. F. FAVREAU, "A Modern Approach to Open-Pit Blast Design and Analysis", *C. I. M. Bulletin*, June 1972.

CHAPTER 14

Special Techniques

"Bulling" or Chambering of Holes

The bottom of a small diameter borehole (such as diamond drill or wagon drill hole) may be enlarged by lighting and dropping a gelatine dynamite safety fuse primer into the hole. The burden is of course, too great for this small charge to break the ground. Its effect is to pulverize the rock in the vicinity and to expel it from the mouth of the hole in the form of dust. Meanwhile, the bottom of the hole has been increased in diameter in the form of a bulbous-shaped chamber, which will then accommodate a larger quantity of explosive. Care should be taken to allow the bottom of the hole to cool down before adding further explosives. The operation is variously called "bulling" or "chambering" or "springing". The method has been used chiefly in quarry work on bench holes.

Tunnel or Coyote Blasting

Where very large quantities of rock are to be blasted and there are adequate free faces (as, for example, on the end of a spur ridge, or the side of a hill), large quantities of explosives can be placed to better effect in excavated chambers rather than in boreholes.

In this way, a large blast can be planned; internal access-ways and chambers can be excavated at one or more horizons, explosives packed into the chambers in their original cases, and the connecting passageways back-filled.

Charges are usually fired with detonating cord.

Perhaps the most notable coyote blast in North America[1] was a well-planned under-water firing of a navigation hazard known as the "Ripple

Rock" in Seymour Narrows, 110 miles north of Vancouver, BC. On April, 5, 1958, 365,000 tons of rock and 320,000 tons of water were displaced with 1378 tons of canned Nitramex 2H blasting agent placed in tunnels and chambers excavated in the submerged rock and fired with primacord. Shafts were sunk in the opposite shores and connected by an under-water drift, from which a raise was developed. From the top of this raise, coyote chambers in the upper part of the submerged rock were then excavated.

Secondary Blasting

Wherever rock is blasted from a face (except possibly in development headings in a mine), some of it may break into pieces too large to be handled satisfactorily through chute openings, into and out of cars, trucks, skips, shovel dippers or grabs, and into crushers.

The equipment in use necessarily limits the size of such pieces. They must be broken to the limiting size, usually by explosives. This procedure is called "secondary blasting" and is achieved either by "blockholing" or "blistering" (see Glossary).

1. **Blockholing:** Short holes 4—6 inches deep are drilled in the boulders. It is not necessary to drill into the centre of a boulder. All that is required is to explode a light charge (about one-quarter of a cartridge, with detonator and fuse) in a shallow hole to crack the boulder. A very large boulder may require several such holes, strategically placed.

However, in built-up areas, it may be *desirable* to drill the hole into the centre of the boulder, to minimize the possibility of damage from flyrock.

2. **Blistering (or plastering):** This method does not involve the use of drilling equipment but of much bigger explosive charges. The charge, suitably primed, is laid on the surface of a boulder and sealed or plastered ("mudcapped") with clay. Ammonia dynamite or packaged AN/FO may be used for this purpose.

Much noise and air concussion results.

Techniques involving the use of shaped charges have been developed by Goex for breaking boulders.

Seismic Blasting

Geophysical prospecting has become very important in the search for deep oil structures and for locating buried metalliferous orebodies. The use of the seismograph is one such important method.

Seismic prospecting depends upon the fact that the earth is elastic. Explosive charges fired in boreholes are reflected and refracted in different ways by abnormalities in the strata, and the time of travel of the various shock waves is determined from signals received by sensitive geophones on reaching the surface.

In order to initiate these shock waves, special purpose explosives have been developed to provide a high degree of reliability, high strength, and excellent water-resistance. To aid in charging deep holes, seismic explosives can be provided with cardboard tubes or threaded cans which can be coupled together to form a continuous charge.

Overbreak Control [2, 3, 4, 5, 6]

Contracts which call for rock excavation commonly contain a penalty clause for unbroken rock left inside a certain gauge line near the nominal perimeter of the excavation; on the other hand where the excavation is to be concrete lined, the contractor may have to pay for the extra concrete required where overbreak has occured. Consequently it is usually important that the walls of the excavation be as smooth and as close to the nominal gauge line as possible.

Similarly, there is a growing need in most rock excavation projects to give greater consideration to the effects of blasting upon the stability of the remaining rock mass (which is, in many cases, seriously affected by uncontrolled blasting programmes).

Based upon these desirable aims, a new era of precision blasting was developed, initially in Sweden, and then by Canadian Industries Ltd. in 1953 during the construction of a hydro-electric facility in Ontario. The technique has since been applied to excavations for highway and railroad cuts, tunnels, dam spillways, canals, and underground machinery halls.

The chief advantages to be derived from the employment of this technique are set out below.

1. The stability and strength of the remaining rock is relatively unaffected.

2. More regular outlines of the remaining rock are preserved thereby contributing to safety and aesthetic appearance (Fig. 48).

3. In built-up areas, less ground vibration is transmitted to neighboring structures.

4. Overbreak is greatly reduced; this may represent a superior economic advantage, by reducing the volume of concrete required as in concrete-lined tunnels.

5. In some tunnelling operations, the need for costly concrete lining procedures may be dispensed with altogether.

6. In underground lode mining, the stability of weak stope walls can be better preserved; and dilution of the ore is minimized.

There are various ways in which blasting may be carried out to achieve these aims, and the bases of these methods are given below. The differences in rock types call for wide variations in methods and drilling/loading parameters used; and fine distinctions between each method at different job

Fig. 48. An example of perimeter blasting on a highway project.

locations have developed a confusion in terminology. In any large or medium scale operation extensive trials should be undertaken to develop the most effective method. The method finally selected will depend upon the nature and economics of the particular operation and the rock characteristics at the site.

All methods involve the drilling of a line of holes along or around the perimeter gauge line, in order to provide a line of weakness or fracture plane ("shear curtain") along which the rock will separate when the main round is fired. This aims to provide a smooth finished surface along the perimeter of the excavation. Similarly, all methods provide for a buffer zone between the perimeter holes and the last or nearest row of the main primary round. The width of this zone (known as the "buffer distance") is usually related to the perimeter hole diameter, but is in any case about one-half of the regular burden distance in the main round.

In practice, when the main round is fired, the buffer zone of rock becomes lightly fractured up to but not beyond the perimeter gauge line. It may then be readily removed with a scaling bar. A smooth finished surface along the perimeter of the excavation will thereby be provided, whether it is a highway cut, building excavation, tunnel or underground machine hall outline or stope wall.

Variations in the method to be selected will be based upon:

1. The rock properties (whether relatively homogeneous, laminated, or intermediate in nature).

2. The conditions of existing ground stress.

3. The diameters of the perimeter holes to be drilled.

4. The spacing between adjacent holes along the perimeter.

5. The "buffer distance" between the perimeter holes and the nearest row of the main round.

6. The diameter of the explosives cartridges to be used.

7. The density of the explosive, and the degree of continuity of the charge.

8. The charge weight per unit length.

9. Whether full de-coupling or a secondary loading of stemming material is used to close the air spaces.

10. The collar stemming to be used.

11. The number of holes to be left uncharged, if any.

12. Whether instantaneous or delayed firing is to be used; and the delay period, if used.

13. Whether fired before or after the main round; or not at all.

The *modus operandi* of overbreak control programmes has been discussed in detail by many scientists, including LANGEFORS[2], GUSTAFSSON[3] and CULVER.[4]

1. Line drilling: In this method regular rockdrill holes, about $1^1/_2$ inches diameter, are spaced 4 to 6 inches apart along the perimeter and left uncharged. Larger diameter holes could be placed at greater spacing intervals. The buffer zone between the perimeter holes and the last row of main round holes should be about 8 to 12 inches wide (the buffer distance). The drilled unloaded holes along the periphery alone provide a plane of weakness to which the rock will break readily and cleanly when the round is fired.

The main disadvantages of Line drilling as compared to other methods of overbreak control are:

1. A greater amount of drilling is involved.

2. Precise hole alignment is required to establish an effective plane of weakness; and

3. The method is applicable only to massive homogeneous rock.

On the other hand, in massive homogeneous rock, and where smooth outlines are specially important, the higher drilling cost of this method may be readily justified.

2. Cushion blasting: This technique is similar to line drilling, with a single row of perimeter holes, but loaded with light decoupled charges. Successful cushion blasting depends upon stemming of all air spaces in the hole, especially when the rock is not homogeneous.

It is therefore applicable to vertical down holes on the surface. One practice is to load the holes by attaching cartridges at intervals to DC downlines which are then lowered into the holes. The holes are completely backfilled with stemming material. To aid the passage of the stemming material down the hole, the DC line may be gently shaken, or raised and lowered as the material is added. Commonly the holes are 2—4 inches diameter for cartridges of $1^1/_4$ inch diameter. Such charges are fired after the main charge, usually on a delay system.

3. **Smooth wall blasting:** When the rock is reasonably competent, smooth wall blasting techniques can be used to advantage in underground applications. Horizontal holes are charged with small-diameter low-density decoupled cartridges strung together and by providing good stemming at the collar of the hole. Charges are fired simultaneously after the lifters. If the rock is incompetent, smooth wall blasting may not be satisfactory.

4. **Presplitting:** Presplitting consists of creating a plane of shear in solid rock on the desired line of break. It is somewhat similar to other methods of obtaining a smoothly finished excavation, but the chief point of difference is that presplitting is carried out before any production blasting, and even in some cases before production drilling.

These methods may be applied to all excavations, both on the surface and underground, where the walls of the excavation must be located within precise limits. Where smooth walls are required in underground excavations such as tunnels and machine halls, the smooth wall effect may be obtained by the use of Xactex, an explosive packed in $^5/_8$ inch diameter cardboard tubes which may be conveniently joined together; when these are loaded into $1^1/_4$—$1^1/_2$ inch diameter holes, a substantial air-space remains into which the gases may expand to reduce the shattering effect on the rock wall. Normally the perimeter holes loaded with Xactex (or similar explosive) are spaced approximately 18 inches apart.

For surface operations, perimeter holes from 2 to 5 inches in diameter at 20 to 40 inch spacing are commonly used.

One problem which has arisen is that rock may be left at the toe. To overcome this, a heavier charge may be loaded at the bottom of the presplitting holes. In addition, the buffer distance may be reduced. For trial purposes, it is suggested that the buffer distance selected be $^1/_3$ to $^1/_2$ the normal burden distance adopted for production blasting.

Presplitting has been carried out in which alternate holes only are loaded, and also in other cases short-delay firing has been used. However, experience has shown that a cleaner face is achieved if every hole in the presplit line is loaded and the holes fired simultaneously, about 50 or more milliseconds before the main round is fired.[5]

Although interest in presplitting is increasing, the number of cases in which details are known is not sufficiently large to give necessary experience from which general charging ratios can be stated. The nature of the rock involved, including hardness, brittleness, and presence or otherwise of cracks, is the most important variable. However, such experience as has been gained to date indicates that a charge ratio of the order of $3/4$—1 oz/ft^2 of area to be presplit might be used in initial trials.

The presplitting technique may be used in all engineering works where smooth faces are required to obtain the advantages of saving in labor and concrete. In addition, the creation of smooth unshattered walls increases the factor of safety since falls of loose rock are much less likely to occur.

The formation of a line of shear may be an advantage where blasting vibrations are a problem, in that the shear line will act as a barrier to the vibrations between the blast area and nearby structures. It may be possible, as a consequence, to increase permissible charge weights per delay, thus decreasing the number of blasts necessary.

Explosive Mining [7]

Under certain limited conditions overburden may be dislodged and directly moved into position in the spoil pile by the planned use of a large quantity of AN/FO loaded into medium diameter holes. The method is known as "explosive mining". Up to 50 per cent of the overburden can be moved in this way, thereby increasing the capacity of the excavating equipment.

As an extension of this technique, Russian explosives engineers have developed the principle of directed rock ejection by methods using conventional explosives. In this way, they have planned and constructed several large canal installations and rockfilled dams. One such dam across the Vakhsh River involved the emplacement of a total weight of 1800 tons of explosives in the banks. From the right bank 1.55 million cubic metres of hard limestone rock were torn out and directionally ejected to form a dam 60 metres high across the river. Then, 30,000 cubic metres of soft earth were thrown into the stream bed from the opposite bank to provide an impervious core. Such a method greatly simplifies dam construction.[8]

Blasting in Plastic Rock Material

By far the largest usage of commercial explosives is for breaking hard brittle elastic rocks. When plastic rock materials, such as stiff clay, mudstone, soft shale, industrial minerals and permafrost are to be blasted, different procedures are called for.

In the former case, the rock immediately surrounding the shot point is pulverized, cracked and weakened by the compressive stress induced by detonation. Following this phase, breakage is readily effected by tensile spalling from the free face, aided by gas pressure (known as "bubble energy").

On the other hand, plastic materials become compacted and strengthened around the shot point when high velocity explosives are used. Tensile spalling is relatively ineffective unless the zone of compaction can first be broken in shear. This would call for greater explosive energy. In such circumstances, it is much better to use explosives of low brisance or de-coupled charges to lessen the initial compacting effect.

With industrial minerals, such as white clays, rock salt, gypsum and the like, the problem is alleviated by first undercutting a kerf with a conventional coal cutter, before firing.

Possible examples include:

(a) The KLADNO method of reverse sequence blasting in subway tunnels in weak slate under the city of Prague, Czechoslovakia.[9]

(b) The findings of Academician MELNIKOV[10] who shows that a certain amount of de-coupling gives better overall blasting efficiency in plastic materials.

(c) Experiences related by SHORT in connection with blasting in rock salt.[11]

(d) Blasting problems in frozen ground.

References

1. C. H. NOREN, "The Ripple Rock Blast", *University of Missouri School of Mines & Metallurgy,* Fourth Annual Symposium on Mining Research, Bulletin No. 97, November 1958.

2. U. LANGEFORS and B. KIHLSTROM, *The Modern Technique of Rock Blasting* (New York, Wiley, 1964).

3. R. GUSTAFSSON, *Swedish Blasting Technique* (Gothenburg, 1973).

4. R. S. CULVER, "Pre-Split Blasting", *The Mines Magazine,* March 1966.

5. Atlas Chemical Industries Inc., "Overbreak Control". Feb. 1967.

6. E. I. DuPont de Nemours & Co., "Four Major Methods of Controlled Blasting", 1964.

7. A. PISANESCHI and H. S. FRAZIER, "Moving Overburden with Explosives", *Mining Congress Journal,* July 1963.

8. R. S. ARIEL *et al.,* "Explosively Constructed Dam of the Baipazinsk Hydro-complex on the Vakhsh River", Translation No. 789, Bureau of Reclamation, Denver, Colo. Oct. 1968.

9. *Engineering News-Record,* October 1, 1970.

10. N. V. MELNIKOV, "Influence of Explosive Charge Design on Results of Blasting", *International Symposium on Mining Research, University of Missouri School of Mines & Metallurgy and U. S. Bureau of Mines,* Feb. 1961.

11. N. M. SHORT, "Fracturing of Rock Salt by a Contained High Explosive", *Colorado School of Mines Quarterly,* Vol. 56, No. 1, 1961.

CHAPTER 15

Cut Holes

For economical blasting, as shown in Chapter 12, it is necessary to have a free face to which the rock can be blasted. In some applications of blasting, no such free face exists. In these cases, such as in tunnels, shafts, and other mine development openings, it is necessary to provide a free face artificially. This is normally done, as a preliminary to the main blast, by drilling and firing an opening in the drilling face. This opening is termed a "cut".

In hard rock, a cut must be fired in each and every cycle of advance. The full round is usually drilled in the one operation; and the complete round is similarly charged and fired in the next operation. But of course the cut holes are the first to be fired in the whole sequence.

Type of Cut Used

The cut should normally be as small as the physical characteristics of the rock will safely permit with a minimum footage of drilling and explosives consumption. But to allow for inaccuracies in drilling and the natural vagaries of the rock, it is usual to be liberal when assessing the quantity of the charge.

The type of cut selected depends not only on the physical characteristics of the rock and the presence of joints and planes of weakness, but also on the skill of the operator, the equipment used, the size of heading and on the desired length of round to be "pulled".

There are five main types of cut (with variations of each). They are:

1. **The centre cut:** This is often called a "pyramid" or "diamond" cut (see Fig. 49). Four (or six) holes converge to a point about 6 inches beyond the depth of the rest of the round.

SIDE ELEVATION

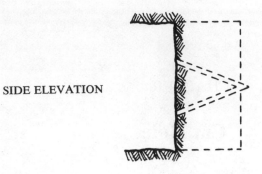

ELEVATION   (easer holes
                    not shown)

ISOMETRIC
VIEW

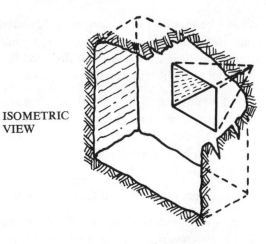

Fig. 49. Centre cut.

This arrangement allows for a concentration of high strength explosive near the apex of the pyramid. By firing these holes simultaneously, a pyramidal opening is obtained which provides free faces into which the other surrounding holes can be fired in turn. Centre cuts are very effective in hard tough rock; but explosive consumption is high and much concussion and flydirt result. A high degree of skill is required to place the holes accurately.

2. **The wedge cut:** This is sometimes referred to as the "Vee" cut. In this case, four (or six) holes converge, not to a point as with the centre cut, but to a line forming a wedge-shaped rather than a pyramidal opening when blasted. It is easier to drill than a centre cut but is less effective in dense tough ground. An example of this type of cut is shown in Figure 50.

For best effects the apex angle of the vee cut should not be less than 60 degrees. This involves a wide spread between the collars of holes. The consequent need to avoid the possibility of fouling the walls with the drilling equipment necessarily limits the depth of round pulled, except for very wide headings such as in tunnels and room-and-pillar faces in hard rock. For this reason, vee cuts are usually restricted to these applications.

3. **The draw cut:** This is like one side of a wedge cut. The holes may be drilled to "drag" from:

(a) The floor, as is usual.

(b) The back (not to recommended as it causes shattering).

(c) One side or other (in which case it is sometimes termed the "fan" cut).

(d) A convenient slip plane or slide.

This cut is suitable for shale or laminated rock. It calls for less accuracy in drilling but requires more holes. It is much less effective than a centre cut in hard ground. Its main use lies in controlled blasting alongside important installations or against timber.

A typical draw cut (off the floor) is shown in Figure 51.

4. **The burn cut:** This cut is suitable for hard, brittle homogeneous rock such as sandstone or igneous rock. It is not so effective in broken or shaly ground. However, its effectiveness can be varied to suit the ground by the use of a selection from a multiplicity of patterns.

Because all holes are normal to the face, longer holes can be drilled and longer rounds "pulled" than with the other cuts, which involve angled drilling.

PLAN

ELEVATION

(easer holes
not shown)

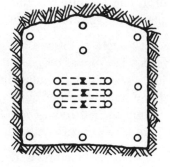

ISOMETRIC
VIEW

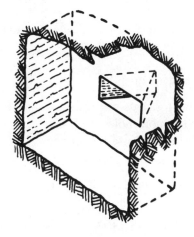

Fig. 50. Wedge cut.

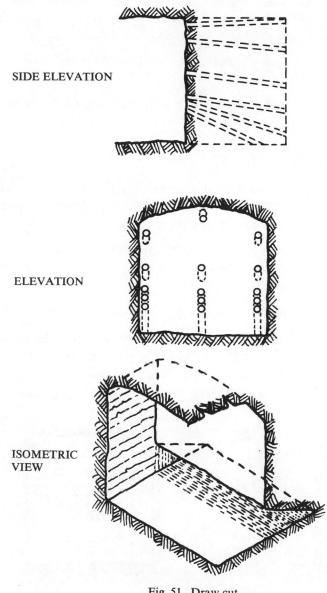

SIDE ELEVATION

ELEVATION

ISOMETRIC
VIEW

Fig. 51. Draw cut.

Certain of the holes are left uncharged in order to provide a series of miniature free faces to relieve stress, to reflect the compressive wave in tension, and to provide open space into which shattered rock from the charged holes can break.

Although there are many variables involved, the main essentials for a satisfactory burn cut (or cylinder cut) are:

(a) Holes must be collared accurately to pattern and must not converge or diverge.

(b) Low power explosives should be used to avoid packing or "freezing" of broken rock in the uncharged holes.

(c) Cut holes should not be fired instantaneously but in sequence, with ample delay periods between consecutive shots in order to give rock fragments an opportunity to be expelled from the cut.

As an aid in planning hole spacing to give "clean blasted holes", a diagram relating hole diameter to distance between hole centres has been developed by LANGEFORS.[1]

Some examples of burn cut patterns are shown in Figures 52, 53, 54.

5. **The large-hole cut:** This type of cut has evolved from the success achieved with burn cuts, in which parallel horizontal holes are used to pull longer rounds, and in which one or more holes are left uncharged.

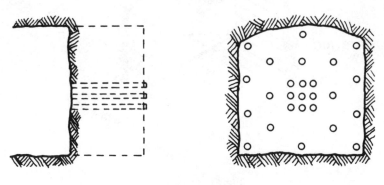

SIDE ELEVATION

ELEVATION
(easers not shown)

Fig. 52. Typical burn cut.

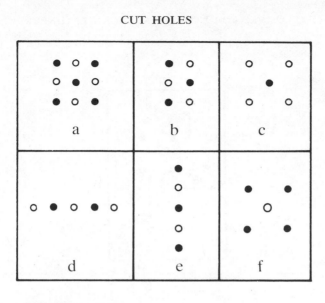

a, b, c = Box cuts
d, e = Line cuts
f = 3 in dia. uncharged hole
● = Charged holes
○ = Uncharged holes

Fig. 53. Burn cut patterns.

Recent improvements in drill-equipment design have made possible the drilling of larger diameter holes. One or two such holes can be left uncharged to provide a more effective miniature free face than the larger number of smaller uncharged ordinary holes used in burn cut patterns. At the same time the need for drilling precision is not so rigid, although still an important feature of any parallel hole cut.

There are several models of large-hole cuts, as set out below.[2]

A **single large-hole cut** is shown in figure 55. The large centre hole is probably of the order of 3 to 5 inches in diameter, with the smaller ordinary holes of $1^{1}/_{4}$ to 2 inches diameter, detonating in the short-delay sequence indicated. Two sets of easer holes are shown, each fired simultaneously in turn at half-second intervals after the cut. In such cases, the burden distance from each cut hole to the wall of the large hole is usually 70 per cent of the large hole diameter. These cuts are suitable for a round depth of about 8 feet. A similar pattern is known as the Fagersta cut.

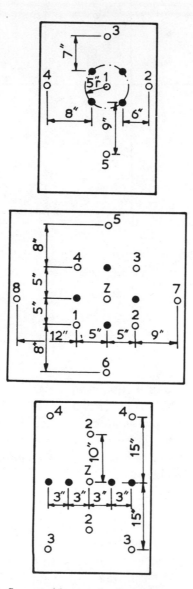

Fig. 54. Suggested layouts for burn cuts (with easers)
Uncharged holes are shown in black.

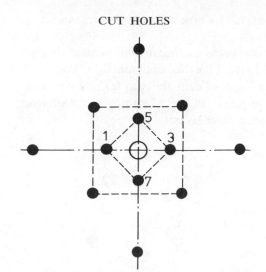

Fig. 55. A single large-hole cut, with two sets of easers.

A double large-hole cut is shown in Figure 56, along with two sets of easer holes. Cut holes are fired sequentially at 50 msec intervals. In this case, the burden distance from the wall of an individual cut hole to the wall of the nearest large hole is normally 0.7 times 2d, where d is the diameter of the large holes. Each set of easer holes is fired simultaneously in turn at half-second intervals after the cut. In Figure 56, where two 3 inch holes are used, the dimensions shown are in inches. S is about 21 inches and B is of the order of $13^{1}/_{2}$ inches.

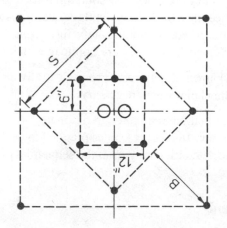

Fig. 56. A double large-hole cut, with two sets of easers.

The **Coromant cut** is of special significance. The holes are located to give maximum effect at the time of detonation. Two overlapping large holes are drilled with a special drill guide to form a slot. The charged holes are drilled from a template. Figure 57 shows the hole layout but does not include the necessary easer holes.

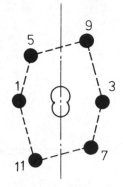

Fig. 57. The Coromant cut.

General Notes on Cut Holes

Cut holes are generally drilled 6 inches deeper than the remainder of the round. This ensures that the other holes leave clean "bottoms" in preparation for drilling the next round.

Stab holes or "baby cut holes" are sometimes used to assist the cut. They are short holes meeting at a shallow depth, and arranged to detonate before the main cut holes. In conventional safety fuse firing, there is a danger of their "cutting off" a cut hole. They are applicable mainly to centre cuts.

Easers (also known as relievers) are used to divide or ease the burden between succeeding holes where necessary.

It is important that holes planned to detonate at a certain instant do not "bottom" within, say, 18 inches of holes planned to detonate earlier. Otherwise, sympathetic detonation may spoil the sequential planning of the round with poor results.

Depth of Rounds

Normal rounds in development headings generally "pull" about 5—6 ft; but burn-cut rounds may pull 8—10 ft.

The economic depth of round pulled depends on the cross-section of the face (or on its smaller dimension) and on the economic depth of shothole of a diameter sufficient to accommodate a standard explosive cartridge in the bottom.

It also depends on the length to which all holes can be drilled in the time allowed for drilling, so that the round can be fired at the end of the shift to maintain an effective cycle of operations.

Another limiting factor to the depth of round pulled is the degree of angling required to drill and fire satisfactorily a centre, wedge, or draw cut. This is of no particular significance in the case of burn cuts where no angling is required. Hence the greater depth of round which can be pulled with a burn cut.[3, 4]

References

1. U. LANGEFORS and B. KIHLSTROM, *The Modern Technique of Rock Blasting* (New York, Wiley, 1964).
2. R. GUSTAFSSON, *Swedish Blasting Technique* (Gothenburg, 1973).
3. R. L. BULLOCK, "Fundamental Research on Burn Cut Drift Rounds", *University of Missouri School of Mines & Metallurgy, Third Annual Symposium on Mining Research*, No. 95, 1958.
4. R. L. SCHMIDT, R. L. MORRELL, D. H. IRBY and R. A. DICK, "Application of Large-hole Burn Cut in Room-and-Pillar Mining", *U. S. B. M. RI 7994*, 1975.

CHAPTER 16

Quarry and Open Cut Blasting Practice

General

The breakage of stone from suitable surface deposits for its manifold uses in the secondary and tertiary industries is of increasing moment on the national scene. By far the largest tonnage of any category of ore or mineral matter produced by the use of explosives by surface methods is in the area of quarrying. For this purpose, a *quarry* may be considered as an open-cut mining operation in which rock, rather than mineral matter, is produced. The mined rock (or "stone") may be used as a raw material feedstock for chemical plants (such as limestone for cement making, and clay shales for brick production); or simply as "construction stone" for concrete aggregate, highway pavements or railway ballast. For this latter purpose, the rock as mined is crushed and screened to provide several graded products for specific usage requirements. For many of these market outlets the rock must meet standard specifications in relation to physical and other characteristics. On the other hand, special methods are involved in the quarrying of dimension stone (building and monumental stone) to produce blocks of rock free from flaws or planes of weakness.

Explosives are necessary for the breaking of stone, the charges being placed in such positions within the rock that their energy, mainly in the form of high pressure gases, will be utilized to the full. However, it is necessary to have equipment capable of drilling holes in such positions and to such depths as will enable the explosives to be placed in the best positions for effective results. Explosives and drilling are therefore complementary in the breaking of stone, and in any discussions on quarrying, explosives practices and drilling techniques are of prime importance.

The modern concept of quarrying is that, as far as possible, the rock should be broken to the required size in the primary blast, with a minimal need for secondary breaking of oversize boulders. This calls for good fragmentation.

A high face (see Figure 58) of rock or ore for quarrying or open-cut work is normally divided into several steps or benches for ease and safety of operation. The width of bench should be sufficient to accommodate the spread of rock broken from the face and to provide space to deploy shovels, haulage units, and drilling equipment. The height of an individual bench depends on the depth capacity of the drilling equipment, and also on the degree of fragmentation required on the particular rock. (Some bench heights are necessarily determined by the angle of stratification of the rock and by the presence of clay seams or planes of weakness.) Most large scale operations use a nominal bench height of 40 to 50 ft. Higher benches (deeper holes) generally mean coarser fragmentation, reduced accuracy in locating the hole bottom, but fewer drill set-ups. Lower bench heights are more suitable for less sophisticated drilling equipment, but drill set-up time and cost is considerably increased.

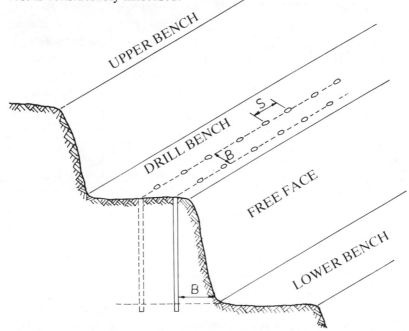

Fig. 58. Cross-section of quarry (open cut) bench system showing typical drilling pattern geometry for two rows of holes.

B = burden distance

S = spacing distance

Figure 58 shows one typical arrangement of benches for quarry or open-cut work. Blast holes are normally drilled vertically, or at a steep angle, in rows parallel to the face. Alternate rows of holes may or may not be staggered.

In order to avoid the development of a toe, and to achieve good fragmentation, special care is taken to site accurately the collars of vertical blast holes. Also, in order to ensure the required depth of subgrade drilling, the reduced level of each hole collar needs to be observed. These details are of increased importance with angled holes because of the greater difficulty in ensuring that the bottom of the hole, which is the most important part of the hole for effective blasting, is at the planned point within the rock. Factors that are involved are, (a) the siting of the collar; (b) the direction or azimuth of the line of the inclined hole; and (c) the extent of possible hole deviation.

For these reasons, some operators find that it pays to make routine surveys and to peg and plan every bench round drilled. This aims to ensure regularity in primary blasting, free from toe, and to achieve a low powder factor, with good fragmentation.

Where the rock is of such a variable nature that random toes form unpredictably, cautious operators prefer to drill only one row, or perhaps only two rows, of holes for a single blast round. On the other hand, in rock subject to back-break, where the bench surface near the new face line is cracked or disturbed, it is difficult to collar holes satisfactorily. For this reason, some operators prefer to drill and fire three or four rows of holes so that a greater proportion of the total number of holes can be collared and loaded in solid ground.

Figure 59 is a diagram displaying one typical firing order for short-delay blasting of a quarry or open-cut face, where three rows of holes are employed in staggered formation.

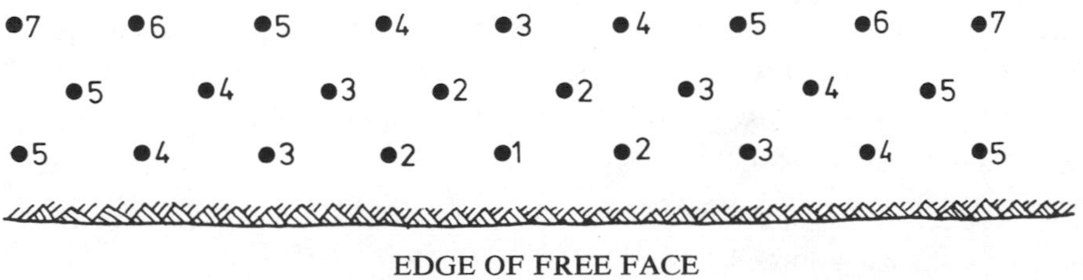

EDGE OF FREE FACE

Fig. 59. Typical firing order for a staggered 3-row pattern on a quarry/open cut bench, using short-delay detonators.

Similar techniques to quarry operations apply also to those of open-cut mining of ores and industrial minerals.[1] In both cases, the prior or concurrent removal of the overlying worthless rock material, the *overburden,* may call for the use of explosives. The overburden may be a relatively thin cover of weathered earthy material in which case it can be ripped and removed by bulldozers and tractor-drawn scoops. But where it is well consolidated, it will need to be loosened or fragmented by explosives before removal. For deeper deposits of rocky overburden or where large amounts of rock along the hanging wall of deep open-cuts are involved, the overburden will need to be quarried in benches, similar to the procedures involved in the mining of ore.

Where near-surface deposits of coal are to be strip-mined, weathered overburden, up to 100 ft deep, is first removed by large-scale excavators, in order to expose the coal seam. In many cases, the overburden needs to be blasted or loosened by explosives, to enable the excavator to perform effectively.

The batter angle of a face of rock or ore depends largely upon the class of rock and its geological structure. Some rocks such as basalt or stratified limestone safely stand almost vertically. Other formations need to be battered at a safe slope angle to avoid local landslides, especially if influenced by the action of water following heavy rains. Under these conditions, slope stability can be extremely tenuous. Yet when barren overburden overlying an ore deposit is to be mined, every additional degree of slope angle gained represents considerable savings in quantities mined and therefore in cost dollars.

The technique of explosives usage in quarrying depends to a large extent on the drilling procedures employed, and therefore on the capital and operating scale of the operation, and the drilling equipment already available on the site. Small scale operators with limited capital investment will necessarily be restricted to simple units of equipment. A significant proportion of the construction stone output is contributed by many small producers using equipment of limited capabilities. Some coverage is therefore given here to their field of interest.

The following analysis of explosives usage in quarries is based upon the drilling practice adopted.

Drilling

Most drills for primary blast holes in surface mining and quarrying are designed for vertical "down" holes, although some now have the capability for angled holes. Manufacturers are presently paying greater attention to devices for lining-up and inclining the holes with greater accuracy. Rigs for surface blasting can drill holes from 2 to 15 inches in diameter and from 30 to 60 ft deep. They use either rotary or percussive tools, with either water or dry (compressed air) flushing systems. They are mostly mobile (self-propelled) and provided with diesel or diesel-electric power units, operating through hydraulic systems. Modern units incorporate fast drilling, quick set-up, and rapid hoisting and lowering speeds, some with automatic drill rod handling.

1. **Rotary drills** are the more versatile. In soft material, drag bits can be used. For medium to hard rocks, tri-cone roller-cutter bits with a variety of tooth facings are used, with compressed-air flushing. These are generally capable of drilling 6 to 9 inch diameter holes (with some units up to 12 inches). Some small truck-mounted rigs use a drag bit, with augers to remove the cuttings.

2. **Percussive drills** break hard rock by high-energy blows delivered by a reciprocating high-speed piston to the drill steel provided with a chisel or button- faced bit. The piston is operated by compressed air under 100 to 200 lb/in^2 pressure. The hollow drill steel is rotated either by a rifle bar mechanism or by an independent rotation device. The hole is flushed by an air stream passing through the hollow drill rods.

There are two main types of percussive drills for surface drilling of blast holes, (a) an above-surface machine which imparts energy to the bit, with the piston blow transmitted through the series of sectionalized drill rods, for drill holes 2 to 6 inches diameter in hard rock, and (b) "down-the-hole" drills, in which the machine follows the bit down the hole, with consequent reduced loss of blow energy because the piston strikes directly on the back of the bit. Down-the-hole drills have their widest application in drilling 4 to 8 inches diameter holes in hard rock.[2]

The smaller variety of surface type percussive drill is mounted on a towed wheeled chassis and termed a "wagon drill", restricted to drill holes from 2 to 3 inches diameter to a depth of 30 ft. Track-mounted self-propelled drills were developed from wagon drills, and have been adapted to serve the large range of surface and down-the-hole percussive drilling needs currently used in the surface blasting of hard rock. Some of these units are truck-mounted for rapid site-changing between widely distant sites.

Jackhammers are percussive rockdrills, operated by compressed air, and typical of those used in underground hard rock mines, but adapted for manual (hand-held) operation. In quarries, they are limited to 1³/₄ inch diameter holes for a maximum depth of 20 ft. Hence they are used only by small operators.

Quarrying Using Small Diameter Holes

In the past, many small scale operators used small diameter holes followed by bulling the hole bottom to accommodate a sufficient explosive charge to break the toe. However, bulling is not now practised to any extent; many now aim to drill holes to a smaller bench height to secure improved fragmentation and greater output. However, this is still not so feasible with jackhammer holes.

1. Short holes by jackhammer: Short hole drilling with jackhammers has been in use for many years; however, it is probably true to say that drilling and firing of these short holes in groups with definite spacings and burdens is a more recent practice. It used to be common, in small quarries for instance, to use safety fuse for initiation of the primary blast. Modern practice however is to drill holes in definite relation one to another and to fire them together electrically or by detonating cord.

Although jackhammer holes may be drilled to greater depth, the usual practice is to limit them to 15—18 ft. At these depths, problems connected with the handling of drill steel, removal of drill cuttings, etc. are not too great.

Faces of stone in excess of 18 ft, however, may be broken by short holes by employing a combination of vertical, inclined, and horizontal holes.

If the convenient maximum depth of holes by jackhammer is taken as 18 ft, faces of stone say 22 ft high may be broken, using a combination of vertical and horizontal holes as shown in Figure 60.

To drill the horizontal toe holes, a pusher air-leg machine may be employed with advantage.

Using an additional line of holes in the face—frequently referred to as half-uppers—permits a face say 27 ft high to be broken by such methods (see Fig. 61). Easy breaking ground and closer than usual spacing of holes would enable this height to be exceeded, as would of course deeper vertical drilling.

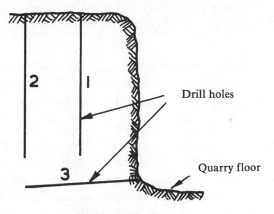

Fig. 60. Typical drilling diagram for 22 ft quarry faces using jackhammer holes.

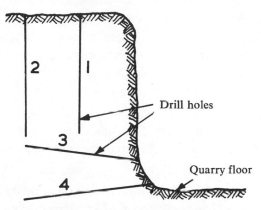

Fig. 61. Typical drilling diagram for 27 ft quarry faces using jackhammer holes.

2. **Short holes by wagon drill:** Perhaps the machine more commonly used for drilling in the manner described previously is the wagon drill. As this machine will, in average ground, comfortably drill 20—24 ft, and to greater depths using special techniques, relatively high faces may be broken employing the foregoing arrangements of holes. Depending on conditions, wagon drill holes could be expected to break quarry faces up to 40 ft high, providing 2-inch diameter explosives are employed.

An important objection to methods which employ a combination of such holes is that the quarry floor, including the toe, must be cleaned up before drilling of the bottom holes can commence. In addition, particularly with high faces, there is the hazard that loose rocks may fall on the men drilling below. It is essential, therefore, to clean the face before drilling commences.

Short hole drilling by wagon drill is usefully employed by the smaller operators who find it difficult to justify large scale equipment. Such wagon drills, equipped with drifter mountings, are capable of drilling 2 to 3 inch diameter holes up to 35 ft deep.

3. **Long holes by diamond drill:** Diamond drilling methods enable a high quarry face to be broken, using small diameter vertical holes only. As the diameter of the drill holes is usually $1^1/_4$ inch approximately, the holes must be placed close together and this results in good fragmentation. Holes $1^1/_4$ inch diameter drilled to break a reasonably high face are frequently located at 5 ft burden × 5 ft or 6 ft spacing.

The diamond drill method, which enables a high face to be drilled from top to bottom using a small diameter hole, gives good fragmentation; but it is now being superseded by more economic methods of drilling.

Quarrying with Large Diameter Holes [1]

In recent years, improved technology and rising labor costs have predicated larger unit sizes of drilling, blasting, loading, transport, crushing, screening, dust recovery, and storage equipment and facilities.

This trend is clearly marked, especially for the medium and large scale quarry and open cut organizations. Mobile-electric or diesel-electric rotary drills are capable of drilling holes from 6 to 12 inches in diameter for depths up to 100 ft, at rates of 400 to 1000 ft per shift, depending upon the hardness of the rock or ore. These drills incorporate air compressors for flushing the cuttings. Drill masts are capable of inclination up to 30 degrees from the vertical.

Either AN/FO or aluminized AN/FO or slurries are used as the primary charge, bottom-primed with high velocity boosters.

Bench heights range from 25 to 50 ft. Holes are usually drilled below grade for distances variously quoted from 10 per cent of bench height, ten times the hole diameter, or generally up to ten feet. This is done to avoid the formation of a toe. With inclined drilling, this is not necessary; and if jet piercing is used, the bottom of the hole is usually chambered to achieve the same objective. Some large open pits use inclined holes for much of their drilling. This is not applicable to the jet piercing process.

Where a quarry is set up to handle large tonnages per day, loading arrangements on the quarry floor and the size of the primary crusher usually permit the rock to be broken larger in the primary blast than would be desirable in a quarry with small-size equipment.

This would allow holes to be spaced further apart than that indicated by the calculated burden distance. Such a procedure is indicated where short-delay firing methods are employed. However, where the ore or rock formation is blocky, fragmentation is likely to be poor and this procedure is contra-indicated.

In the best of open cut operations, some secondary breaking is always necessary. Operators generally aim to minimize the need for secondary blasting, however, since its direct and indirect effect on total costs is exceedingly high.

The iron ore mines of Minnesota generally aim to keep the proportion of secondary breakage to within two per cent of their total tonnage; and then to use a drop ball, in various weights up to 7 tons, handled by a mobile crane, in order to avoid the cost of explosives for secondary breaking. Some of these mines break up to a million tons of ore in a single multiple-row primary blast.

The type of explosive employed and the arrangement of charges in large diameter holes depends on the type of rock, stratification (if any), depth of holes, etc. The depth of hole is important since this determines whether the charges within the hole should be decked or not. Where deck charges are employed, from half to two-thirds of the calculated charge is placed in the bottom of the hole with the balance in one or more deck charges separated by sand.

For breaking softer formations, especially where there is no serious water problem, the use of a low-density blasting agent such as AN/FO avoids the need to deck-charge.

For wet holes in the harder formations, metallized AN slurries are particularly applicable, especially where a pump truck service is available.[3]

References

1. E. P. PFLEIDER, Ed., *Surface Mining* (New York, AIME, 1968).
2. S. TANDANAND, "Principles of Drilling", in *SME Mining Engineering Handbook* (New York, AIME, 1973) Chapter 11.
3. J. DANNENBERG, "Bulk Loading and Explosives Selection", *Roads and Streets,* January 1968.

CHAPTER 17

General Industrial Applications

Surface Excavations

Apart from quarries and open cuts, all other surface excavations such as canals, trenches, cuttings, and foundation excavations are included in this category.

Canals are used chiefly for navigation, irrigation and drainage purposes, and generally as aqueducts; trenches are usually prepared as drains and for laying pipe lines or cables; cuttings are for road and railway formations; foundation excavations for city buildings, structural and machinery installations; and general excavations are required for clearing sites, for pole holes, and for miscellaneous purposes.

In many of these applications, blasting is advantageous to loosen the rock before its removal by mechanical excavating plant. A wide range of rock types and strengths is encountered in this class of work. Soft overburden is first removed to expose the solid rock.

1. **Canals and trenches in rock:** It is necessary to use explosives to loosen the rock along the trench profile in advance of mechanical trench excavators or backhoes. Mobile high speed drill rigs are used.

For narrow pipeline and cable trenches up to 2 ft wide, a single row of holes is drilled to grade level. Spacing is determined experimentally in conjunction with the type and grade and quantity of explosive selected. Holes are fired sequentially towards a free face with short-delay detonators.

In harder ground or with wider trenches up to six feet wide, two or more rows of holes are drilled, sometimes in a staggered formation so that limiting side holes are on alternate sides.

Holes are placed vertically as shown in figure but may also be slightly inclined.[1] The required amount of side batter to the trench is usually achieved by the incidental amount of side overbreak that results.

The explosive charge should be ample but well distributed up the hole at a low loading density to avoid excessive overbreak. Drill holes need to be closely spaced in these "tight" narrow faces. Holes should be well stemmed. Figure 62 shows a typical hole layout in a staggered pattern.

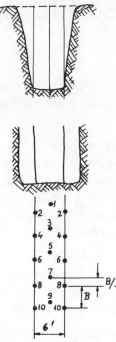

Fig. 62. Diagram for blasting trenches in hard rock, with leading centre hole and two flank holes.

Where important installations must be protected, more holes are used, each charged more lightly, and barricades and blasting mats are used.

2. **Drainage ditches in swamp lands:** For draining swamps, it is often impossible to use either hand or mechanical methods.

The method of ditch blasting with explosives, however, is easy to apply, rapid in results, and, furthermore, requires no elaborate equipment. Channels ranging in width from a few feet to many yards across may be successfully blasted in this manner. The depth of the drain may also be controlled to some extent by the amount of charge used and the depth to which the charge is placed.

Ditch blasting in its simplest form is illustrated by the single line of charges placed along the centre line of the proposed ditch or channel. The charges are placed in holes in the mud or silt at regular intervals; the top of the charge is placed 8—12 inches below ground level. The amount of charge or number of cartridges in each hole will depend on the character-istics of the ground and the width and depth of ditch required. This is largely a matter for experiment under the particular conditions at the site.

As a guide, however, a ditch approximately 4 ft across the top and 3 ft deep can be blasted in silt by charges of one cartridge of ammonia gelatine 60% placed at intervals of 2 ft—the charges being fired electrically or by detonating cord. Charges of $1^1/_2$ cartridges placed as a column in similar type soil will form a ditch approximately 6 ft wide and 4 ft deep (see Fig. 63).

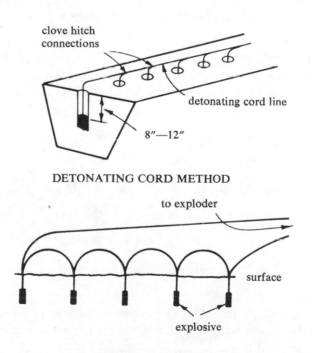

DETONATING CORD METHOD

ELECTRICAL METHOD

Fig. 63. Single line ditching.

For wider ditches the number of cartridges per hole should be increased, still, however, keeping the top of the top cartridge 8—12 inches below the surface level. The depth of the ditch will also be increased in proportion.

When ditches are required to be wider than can conveniently be blasted with a single line of charges (or where a wide shallow channel is required), multiple rows of holes, staggered in relation to one another, are employed. Generally however, the number of rows so employed does not exceed three (see Fig. 64).

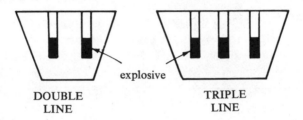

Fig. 64. Double and triple line ditching.

A cross-row method, in which rows of holes are placed at right angles to the main line of charges at intervals along the centre line of channel, is also a widely used method, particularly where wide ditches are required.

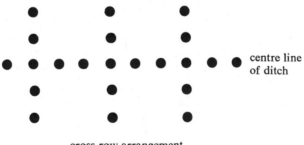

If it is possible to obtain the desired effects with the single line of charges, this should be used, as the method is simpler to operate and is less costly in materials.

In ditching with explosives, detonating cord or the electrical method of initiation is employed. However, since the capacity of the generator-type exploder would limit each firing to a short length of ditch, a condenser-discharge type of exploder should be used.

Detonating cord firing also enables long lengths of ditch to be fired at the one time. The DC lines with primer cartridges attached may be prepared on solid ground near the site and placed in position when ready. This is the method of initiation normally recommended for ditching with explosives. For further details on ditch blasting, see Chapter 21.

3. **Road and railway cuttings in rock:** After removing soft overburden, it is necessary to establish a free face across the width of the cutting. Where a natural free face is not available, it may be established by means of a box cut.

The diameter of cartridge (where dynamite is to be used), burden distance, and hole spacing depend upon the diameter of hole drilled, which in turn depends upon the equipment available.

It is usual to drill holes to about 2 ft below grade level and to use electric short-delay blasting or DC and MS delay connectors arranged so that each row of holes is fired in sequence.

Under average rock conditions, explosive consumption works out at a minimum of $0 \cdot 5$ lb/yd^3 of rock in place.

4. **General excavations:** Based on the general principles expounded above, a specific blasting pattern can be evolved for any type of excavation.

Much depends on the condition and requirements associated with each particular application.

Submarine Blasting

Applications of submarine blasting for civil purposes include the following:

1. Blasting trenches across rivers to carry pipelines and cables.

2. Demolition of wrecks.

3. Cutting of piles.

4. Blasting channels through reefs, bars, and sandbanks.

5. Harbor development and improvement.

6. Blasting sheet piling and coffer dams.

Special provisions relating to submarine blasting include the following:

1. The effect of shockwaves transmitted through the water to nearby installations.

2. The effect of hydrostatic pressure.

3. Special waterproofing requirements and the need for high resistance to water of the explosive components.

4. The difficulty of effective placement.

5. Because of water pressure, burden distances must be reduced.

6. Holes should be drilled well below grade.

7. In order to avoid silting up, charges should be fired as far as possible in one operation.

8. Full advantage of an ebbing tide should be taken to scour out the broken material.

9. Use either Blasting Gelatine or other high strength gelatine with aluminium-tube electric detonators or with DC.

Where a small quantity of submerged rock is to be removed under water, it is sometimes economical to employ a diver to drill holes with a jackhammer and to charge and connect the holes for subsequent firing from the surface. With larger jobs such as deepening or construction of a harbor, special drilling equipment mounted on a drill barge designed for the purpose is generally necessary.

The drill barge is equipped with spuds that can be lowered to the seabed. The spuds, which usually consist of heavy timber, reinforced with steel plate, may be raised or lowered with steam- or compressed-air-driven hoists by means of steel-wire ropes through top and bottom sheaves. Before dropping the spuds, the barge is manoeuvred and secured in the desired drilling position by anchors or spring lines. The barge is then raised a few inches on the spuds to ensure a stable and level drilling platform.

The drilling equipment usually consists of one or more drilling masts mounted on one side of the barge. These are generally arranged on rail tracks for ease in manipulation when "spotting" drill holes. The masts are usually high enough to accommodate the longest length of drill steel required for the deepest hole without stopping to make extensions or adjustments.

The masts also support the sand pipe which is raised or lowered on guides by a hoist and cable. The sand pipe, which is "belled" at the top, is slightly larger in diameter than the gauge of bit used; it is usually of sufficient length to reach from surface to bedrock. For shallow water where wave and tidal conditions are moderate, the sand pipe can be in one piece; but when employed in deeper water in more difficult conditions, a telescopic type is used. Where drills are used under water, the sand pipe used is shorter but should extend from the bedrock through the soft unconsolidated material on the seafloor.

The purpose of the sand pipe is to prevent loose material from falling into the drill hole, to guide the drill steel, and also to convey the drill cuttings away from the hole. In some cases the sand pipe is slotted near the bottom to discharge the heavier cuttings.

Holes drilled for underwater blasting vary from $2^{1}/_{2}$—6 inches diameter, using jackhammers and down-the-hole drills.

The drilling pattern must be so designed that there is no risk of unbroken rock being left above the grade level, otherwise further work becomes difficult and costly. Each drill hole charge will create a coneshaped crater; therefore the holes must be spaced and drilled to such a depth below grade that the craters will overlap. The spacing is usually of the order of 5—10 ft, depending on hole diameter, type and thickness of rock to be removed, and depth of water. Holes are usually drilled below grade to a distance corresponding to the spacing distance.

Where a vibration problem exists, small diameter holes and smaller charges should be used.

The quantity of explosives required will vary according to the depth of water and tenacity of the rock. Usually 1—5 lb are required per cubic yd of rock. As a guide, 2 lb/yd³ should be suitable for reasonably hard rock under 30—40 ft of water. Preliminary examination and survey by diver and sounding is usually required before determining drilling patterns, charge quantities, and methods.

In work involving extensive underwater blasting, special explosives should be used because with increasing depth of water the sensitivity and VOD of the explosive decreases. This effect is probably due to the minute air bubbles within the cartridges being squeezed out, thus increasing the density. To counter this effect, special explosives formulated to give a high

VOD and greater sensitivity are recommended. No special priming is necessary, except that a No. 8 aluminium electric detonator is required.

Explosives recommended for underwater blasting include the high velocity gelatines. Seismic cartridges are supplied in tubes which may be screwed together by means of an outer coupler to form a rigid column of explosive of any practical length. This is convenient for charging readily through a sand pipe or charging tube.

Where the sand pipe reaches to the surface, the charge of explosive may be charged through the pipe and pushed to the bottom of the hole by means of a long tamping stick.

A conventional brass loading tube may also be employed. This should fit inside the sand pipe and penetrate 2 ft or so within the drill hole. The explosive is then pushed to the bottom of the drill hole.

Explosives may also be loaded into metal shells such as stove pipe and the unit of the required length lowered into the drill hole.

A small float which will readily pass through the sand pipe is often attached to the detonator leg wires or the DC so that when the sand pipe or charging tube is withdrawn the ends float to the surface. Connections for firing a blast may be made by using a small boat. For firing electrically, it is important to tape all joints carefully.

When firing small charges, the drill barge need not be moved from the drilling position. However, for multiple holes with larger charges, it should be moved to a safe position.

Vibration and concussion from underwater blasting may be a cause of concern when blasting near harbor installations. If delay firing is indicated, DC lines should be connected to the charges with delay detonators at the surface. It is possible, however, that the underwater charges, being placed at close spacing may become subject to sympathetic detonation, in which case some or all of the charges will explode together. Because of the close spacing of holes in underwater blasting, cut-off holes could result from sequential firing unless rock and other conditions happen to be favorable.

More recent advances in sub-marine blasting techniques have been developed with the use of shaped charges (see Chapter 10).

Pipeline and cable trenches through rock and reefs can be effectively and expeditiously executed without the need for the drilling of holes.

Shaped charges (Fig. 65) are merely placed on the top of the seabed, along the projected trench lines and hooked up to the firing point with detonating cord.

Fig. 65. Shaped charge for under-water ditch blasting

This technique has been developed by the Ocean Applications Group of Jet Research Center, Inc. of Arlington, TX. In the same way they have developed cutting harnesses in which small shaped charges are assembled as circular girdles to cut piles made variously from steel tube (hollow or concrete-filled), reinforced concrete, or timber. These girdles are particularly useful for removal of hazards to shipping, such as the demolition of wrecked off-shore drilling rigs (see Fig. 66).

Wooden piles may also be cut at river or sea bed level by means of a charge of strong gelatine strung in the form of a girdle in a plastic tube, or alternatively, taped to rope, wire, or detonating cord.

In many cases, a diver is employed to place the charges in position in close contact with the pile, especially where heavy marine growth makes this otherwise difficult (and may, incidentally, indicate the need for increasing the charge).

Where a diver is not available, an elastic material may be used to form a girdle around the pile just below water level; the girdle may then be located by sliding it down the pile to the seabed, using two light wooden poles on opposite sides of the pile.

As a guide, a pile of 18 inches diameter would require a charge of 2—3 lb of high velocity gelatine dynamite.

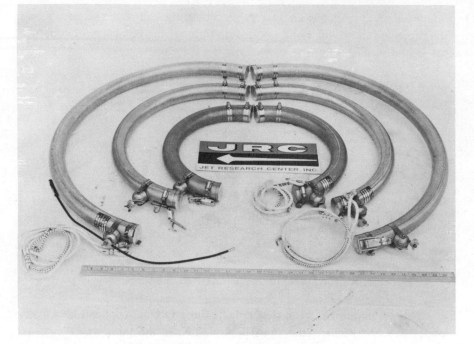

Fig. 66. Outside circular girdles for cutting piles
(shaped charge principle).

Demolition Work [1, 2]

Explosives may be used to demolish a wide variety of materials (including stone, brick, concrete, steel, and timber) in the following applications:

1. Breaking up structures fabricated from steel plate and girders; breaking heavy castings such as flywheels, engine beds, and the like; breaking up iron acid pots and retorts by concussion charges.

2. Demolition of factory chimney stacks.

3. Demolition of masonry and concrete walls.

4. Removing old concrete foundations or adjusting existing foundations to accommodate new machinery.

Generally, each demolition job is an individual problem, requiring special treatment; consequently, this class of work should be placed in the hands of experienced contractors. Because of the complexities of and risks involved in each individual task, amateurs and enthusiastic revolutionaries are likely to become doomed in time to a process of self-elimination. Therefore, in their own interests, no further notes on the methods to be used are offered here.

References

1. R. GUSTAFSSON, *Swedish Blasting Technique* (Gothenburg, 1973).
2. American National Standards Institute, A 10. 6 — 1969.

CHAPTER 18

Blasting in Tunnels and Underground Development Headings

General

This application of blasting is confined to shaftsinking, tunnelling, drifting, crosscutting, raising, and winzing.

It is the most difficult of blasting techniques because (apart from bench cuts in shafts) no free face exists and the firing of cuts becomes necessary. For uniformly good results, burden distances are shortened and explosive usage is increased. The work of drilling, charging, and blasting demands a much higher order of skill than where free faces exist.

The hole with the least resistance is always the next to be fired. It is usually the middle hole of a row. When fired, additional free faces are then available to the remaining holes in turn. This principle is basic to the design of all round patterns. The lifters are always fired last so as to reduce the compaction of the muck pile for easier shovelling, whether by hand or machine.

Tunnels

Tunnel driving is a common type of construction project in civil engineering contract work. Depending upon the class of ground to be traversed, tunnels may be excavated in:

(a) Soft wet unconsolidated ooze, wherein tunnelling shields, with or without the use of plenum systems, are used, followed by tubbed linings (of cast iron or prefabricated concrete segments) for permanent support of the ground.

(b) Soft to medium rock in which tunnelling moles can be used to advantage.

(c) Hard rock, in which the conventional drill/blast/muck cycle is necessary. This is the only type that calls for the use of explosives.

Tunnels are used variously for railways and highways in mountainous terrain and under rivers; for river diversions in dam construction work; for urban underground rapid transit facilities; for hydro-electric and other aqueducts; and for many other miscellaneous applications, including mining.

Tunnels are excavated in cross-sectional dimensions ranging from 10 to 50 ft in diameter and up to 10 miles long. The larger sections are shaped more like a horse-shoe, with a horizontal base and an arched roof. In these large tunnel headings, sophisticated items of equipment such as drill jumbos and Conway shovels are used.

Depending chiefly on their cross-sectional dimensions, tunnels can be advanced "full face", or by pilot heading and subsequently stripping to full face dimensions. See Figure 67.

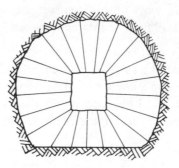

Fig. 67. Large tunnel heading excavated by pilot heading (approx 10 × 10 ft) and by ring-drilling in a fan formation from the pilot heading.

In general, the principles are the same with full face and pilot heading, although with the latter the dimensions are restricted to those of a drift or crosscut.

Various types of cuts are employed, but usually a vee cut.

Where a full face is advanced, drilling is normally done from a hydraulic jumbo. With this equipment, a large borehole (2^1/$_2$ to 5 inches diameter) is sometimes used as the basis for a cut.

Tunnel driving, like shaftsinking, is an extremely costly operation and therefore many benefits and economies can result from the application of modern work-study principles. Well-planned operation cycles are therefore a *sine qua non* of modern tunnelling and shaftsinking operations.

Shafts

Shafts for mine development purposes are usually vertical and may be circular, elliptical, square, or rectangular in cross-section. For many reasons, the circular section is now regarded as the most effective shape. As with tunnel construction, many shafts are now being sunk by independent contractors.

Generally, irrespective of the cross-sectional shape of the shaft, centre or burn cuts are used; otherwise, bench cuts (a special case of the draw cut), used alternately on either side of the face, give good results, especially with hand-mucking in rectangular shafts.

Centre cut: This has the disadvantage that the excessive fly-rock is liable to damage the timber in rectangular shafts. Apart from this, there is the ever present risk (in wet shaft faces) of a misfire. Misfires in centre cuts usually mean the loss of the round. Misfires in wet shaft faces are usually hazardous and difficult to resolve. Shaft faces are usually wet.

Burn cut: These are to be preferred to centre cuts even though the final shape of the muck pile suits mechanical mucking in both instances.

The damaging effect of fly-rock is much less with burn cuts. With either cut the selection of a suitable short-delay blasting pattern gives control of the final muck pile.

Wedge cut: These are generally not favored because of the poor fragmentation effect of the cut itself.

Bench cut or sump cut: This method is favored where handmucking is practised. Half the face is fired at a time in such a way that a stepped or benched effect is maintained, as in Figure 68.

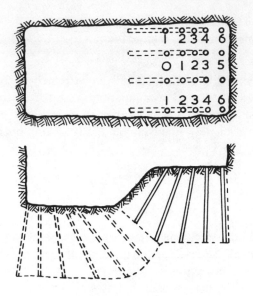

Fig. 68. Bench cut in rectangular shaft.

The particular advantages of this method are:

1. A partial free face is always preserved so that draw cutting is not severe.

2. A sump is always available for drainage. This leaves the bench dry for drilling, charging, and firing.

3. It is an easier pattern to drill, and inaccuracies are less important.

4. There is better control of fly-rock.

5. Charges can be kept dry with less risk of a misfire.

6. The muck is piled high against one end of the shaft. This allows gravity-aided hand-shovelling into an inclined kibble for most of the muck.

Usually a medium-high strength gelatinous explosive is preferred because of its water resistance.

If multiple fuse firing is employed, the capped fuses are always water-proofed. But because of its greater degree of safety, electric firing is more general.

Short-delay detonators give better fragmentation, higher efficiency, and better control of the muck pile.

Drifts and Crosscuts

These are headings advanced horizontally either along or across the strike of the country respectively. Because a crosscut cuts across the bedding planes, shearing of the rock by blasting is more effective than in drifts.

Holes are mostly drilled horizontally by rockdrills mounted on column and bar, or on air-fed pusher legs. With some cuts, however, angled holes are necessary. Angled drilling calls for greater skill and limits the depth of hole which can be drilled because of the fouling of the equipment on the walls, floor, or roof of the drift.

Draw, wedge, centre, burn and large-hole cuts are variously used, the choice depending to some extent on local conditions.

After the cut has been fired, there follow the cut easers, side holes, knee holes, shoulder holes, back holes, and finally the lifters, each group in sequence (see Fig. 69).

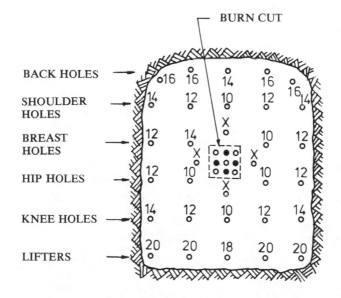

Fig. 69. Typical arrangement of holes in an 8 × 10 ft drift heading in very hard rock, fired with a 9-hole box-pattern burn cut.

If the centre hole of the box cut goes at zero, and the corner holes at 1, 2, 3, 4, then the four easers labelled X could be timed for 5, 6, 7, and 8 delays; or to go simultaneously at 6. The round holes would then be timed for short-delay sequences as shown; or alternatively for half-second delay sequences.

The lifters point slightly downwards to ensure that the floor grade does not become excessive. Similarly, the back holes point upwards to preserve the desired headroom. Back holes are also placed so that an arched "roof" is maintained.

Where much water is to be drained away, the lifter on one side is placed lower in order that a continuous drain or ditch can be carried along with the advancing face.

Apart from purely exploratory openings (which may be smaller), and depending on their particular purpose, drifts and crosscuts in operating metalliferous mines, and pilot headings in tunnels, range variously from 5×7 to 10×10 ft and larger in cross section.

Winzes

Winzes are really small shafts sunk from one level to another (but not including those sunk from the surface). Like shafts, they may be vertical or inclined.

Cross-sectional dimensions are normally 5 by 5 ft, 6 by 4 ft, 8 by 5 ft, and 10 by 6 ft. All types of cuts may be used; in the larger ones, if vertical, the bench method is sometimes used. This aids handmucking into the kibble.

Raises

Raise headings are advanced upwards, in apposition to winzes. Their chief virtue lies in the fact that removal of the broken rock is, due to gravity, virtually automatic. An important disadvantage is concerned with the difficulty of adequately ventilating the heading. Some form of artificial lining is advisable in order to divide the raise so that spoil removal can be localized into one compartment whilst another can be used as an access gangway.

Raises are of a similar range of dimensions to winzes. Centre cuts are seldom used because of the resulting damage to timber. Burn cuts are more generally used.

Air-leg stoper machines are used for drilling from a staging erected 7 ft from the face, or from an Alimak unit; or from a portable cage suspended by a rope through a borehole. However, raises are now commonly excavated by drilling and reaming.

CHAPTER 19

Blasting in Stoping Operations

General

Ore is extracted from orebodies in metalliferous mines by various methods of "stoping". The ore is generally broken from the mass by drilling and blasting. A particular stope is normally delineated and prepared for ore extraction by various headings and/or timber structures appropriate to the particular stoping method selected.[1]

One of the important considerations in stoping is to support the walls of the country rock where necessary on either side of the vein from which the ore is being extracted. In many cases, it is necessary to fill the space resulting from the removal of the ore with waste rock or mill tailing in order to prevent the walls from caving. This filling operation normally proceeds step by step with ore extraction.

It is therefore important that the action of breaking the ore by blasting should not adversely affect the stability of the walls. Designers of stope blasting patterns must therefore bear this in mind.

Apart from this fact, and the necessity to support the back of the stope temporarily while the holes are being drilled, stope blasting is relatively simple. A free face always exists. Holes need only be spaced to suit the burden distance corresponding to the dimensions of the holes drilled. Holes fired near the walls, however, must not be allowed to break back into the walls.

Conventional Stoping

Conventional methods of stoping include horizontal cut-and-fill; inclined cut-and-fill; square-set; open; shrinkage; and combinations of these methods. They each involve the drilling of holes 5 to 8 ft deep normal to the face being advanced, and parallel to the free face. Holes are normally drilled horizontally, inclined, or vertically downwards.

Horizontal (or inclined) breasts or slices about 8 ft high are progressively advanced along the stope (see Fig. 70).

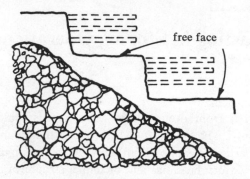

Fig. 70. Overhand stoping with horizontal holes.

The blasting of vertical "up" holes is liable to leave a ragged unstable "back". However, when mechanized cut-and-fill stoping is employed, this problem may be overcome by using hydraulic jumbos, drilling "up" holes inclined at about 60° to the horizontal. In this way, the rock is broken mainly in tension, there is much less shearing, and the need for heavy charges is avoided.

Long-Hole Sub-Level Stoping

With wide massive orebodies justifying a high rate of extraction at low unit costs, it is now the practice to drill a fan-shaped series of long holes from a sub-level drift in parallel vertical planes up to 8 ft apart. The technique is known as "ring-drilling" and the ore is broken by "long-hole blasting" (see Fig. 71).

The average stope block may be 120 ft long and 300 ft high with a lode width varying from 40 to 250 ft or more.

The stope block is first prepared by excavating the usual end raises and the chute raise and draw-point complex immediately above the haulage level. The next major step is to prepare a free face for ring-drill blasting. This may consist of a cut-off stope or "slot", 12 ft wide, excavated by

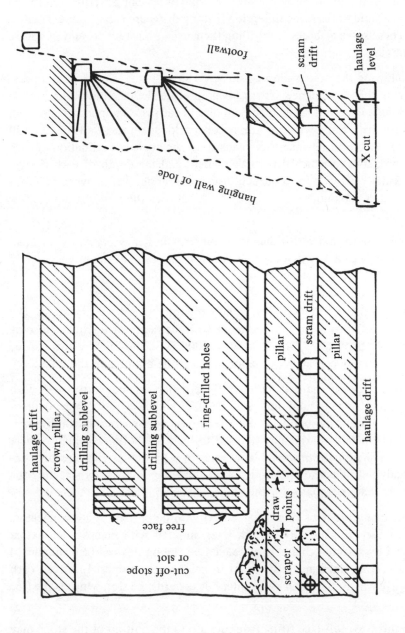

Fig. 71. Longhole stope blasting by ring drilling.

shrinkage stoping across one end of the ore block: or by excavating an opening 12 ft wide from which vertical long holes can be drilled and fired to form a slot. At the same time, ring-drilling drifts are run at sublevel intervals. These serve as bases for drilling the rings of blast holes, and of course for charging operations.

Following the preparation of surveyed stope plans, ring-drilling operations may then commence. These plans indicate the orientation and depth of each hole in a fan-shaped array of holes in each vertical plane (ring) from each drill drift. Surveyors provide collar markings on the walls of the drill drifts for each drill set-up. Stope plans show the angle of inclination of each hole, the depth to which it is to be drilled, and the depth of each charge. Plans usually provide for a burden distance of about 7 ft between adjacent rings, with a maximum toe spacing of 8 ft between the bottoms of adjacent radial holes.

Holes are drilled $2^1/_4$ inches in diameter with heavy percussive drifters mounted in such a way that the holes in each ring can be collared accurately by rotating the drill in its mounting.

One of the problems likely to be encountered is hole deviation. The tendency for holes to deviate increases with the depth of the hole. In some mines, hole depths are limited to 80 ft; otherwise the blast results can be prejudiced by inaccuracies in the locations of the bottoms of the holes.

Holes may be charged with AN/FO, collar-primed with a $1^3/_4$ inch diameter cartridge of ammonia gelatine 60% and electric short-delay detonators; or bottom-primed with non-electric Anodet primer assemblies. Alternatively, a bulk slurry may be used to load the holes, especially where water is a significant problem. The holes in a particular ring are fired sequentially with electric or Anodet short-delay detonators, by selecting the particular delay detonator for each hole.

The firing order of rings is arranged with detonating relay connectors between each ring in a DC trunk line, initiated with electric detonators operated by selected channels of sequential timing devices (see chapter 9) for each set of rings in the stope, or in neighboring stopes. In effect, each successive ring is blasted at delay intervals towards the slot, which is the free face.

In this way, full-face firing from the top to the bottom of the stope may be carried out. The ore broken in each ring with an 8 ft burden may range from a few thousand to as much as 40,000 tons. Broken ore flows through control chutes into scram drifts above the haulage level; ore is then scraped to loading chutes and loaded into trains by gravity. Alternatively, broken ore may be arranged to flow into draw-points and to be loaded by load-haul-dump (LHD) units into ore trains.

In all such cases of primary stope blasting in wide stopes, where there is room to deploy mechanized equipment, the ore is likely to break in large blocks, resulting in poor fragmentation. But the general aim is to break ore from the solid orebody in the primary blast to a size that can be handled effectively through the later phases of the operation. This means that the ore should be reduced in size to suit ore handling activities through ore passes, chutes, chute gates. LHD shovels, into and out of haulage vehicles, and into underground crushers with a minimum of secondary blasting, and at a minimum total cost. This overall problem calls for a continuing in-depth study of all the factors involved, in respect of the characteristics of the particular orebody. The main factors concerned are those of drilling and blasting, but they should not be considered in isolation.

References

1. L. J. THOMAS, *An Introduction to Mining* (Sydney, Hicks Smith, 1973).

CHAPTER 20

Blasting in Coal Mines

Restricted Choice of Explosive

The possible occurrence in underground coal mines of highly inflammable mixtures of methane and air and explosible dusts has led to special statutory requirements for explosives for use in such mines.

Explosives that have passed special tests as set out in Schedule 1-H of the MESA[1] are considered safe for use in coal mines. Only explosives so approved and designated as "permissible" are legally allowed to be used in coal mines in the United States (see Chapter 3).

The main characteristics of a "permissible" explosive are that, when detonated, it must produce a flame

(a) of relatively low temperature,

(b) of small volume, and

(c) of short duration.

Permissible explosives differ then from ordinary high explosives in that they contain an appreciable proportion of one or more constituents whose function is to lower the temperature of the explosion. This has been achieved by the increased use of ammonium nitrate instead of NG; and by the addition of a cooling agent (or flame depressant), such as sodium chloride.

Brand names of currently approved explosives are shown in Chapter 3. The continued permissibility of specific lots of these brand names are contingent upon the following requirements:[2]

(1) The explosive shall conform with the basic specifications, within limits of tolerance prescribed by MESA; the cartridges must be of diameters that have been approved.

(2) The explosive shall be used in conformance with all the provisions of the Federal Coal Mine Health and Safety Act of 1969.

(3) The explosive shall be stored in surface magazines under conditions that help maintain the original product character; it must be used within 48 hours after being taken underground.

(4) The explosive shall remain in its original cartridge wrapper throughout storage and use, without admixture of other substances.

(5) The explosive shall be initiated with a copper or copper-based alloy shell commercial electric detonator (not cap and fuse) of not less than No. 6 strength.

(6) The polyethylene tubular-packaged water-gel permissible explosive shall be handled under the same requirements existing for nitroglycerine-sensitized permissible explosives regardless of the additional safety aspects. Excessive exudation of the water-gel ingredients through broken polyethylene ties or cartridges must be treated as deteriorated explosives.

In Canada, a "permitted" explosive is one approved by the U.S. Bureau of Mines and/or the U.K. Ministry of Energy for use in coal mines, provided it is used in a specified manner, as set out below:[3]

1. The explosive must be in good condition.

2. The size of charge detonated in any shothole shall be within the specific charge limit for that particular explosive.

3. The charge shall be detonated by means of EB caps only; such caps shall be not less than of No. 6 strength.

4. Shotholes which have a burden so heavy they are obviously liable to blow out, must not be charged.

5. Shotholes must not be fired in the presence of a dangerous concentration of combustible gas or dust.

6. Shotholes must not be fired until they have been properly stemmed with an adequate amount of non-combustible stemming.

7. Firing must be carried out only by means of a permitted type of blasting machine of adequate capacity.

In the United States, the tonnage of bituminous coal mined annually is of the order of 650 million tons. Approximately 44 per cent of this tonnage is produced from surface strip mines. Of the remaining 365 million tons won underground, approximately 95 per cent is won by mechanical non-explosive methods. It therefore follows that explosives are presently used to mine 18 million tons of coal annually from underground coal seams, as well as for development work and overburden removal on strip mines.

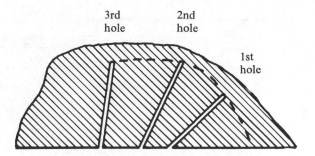

Fig. 72. Grunch blasting.

Coal Blasting

There are two methods of breaking coal with explosives:

1. **Blasting "off the solid" or "grunching":** The first stage in blasting "off the solid" is to form a free face by using explosives; a series of horizontal holes angled into the face in the form of a fan cut is generally used (see Fig. 72). Where pillars are blasted, a free face already exists.

2. **Blasting "cut coal":** It is desirable to provide a free face for any shot; this is achieved by cutting. The normal cut or kerf is 5—6 inches wide and 4—9 ft deep, and is placed either horizontally at floor level, or near the roof, or at some intermediate position. Vertical cuts are also used. The depth and position of the cut depend on the general mining conditions.

The position of shot-holes for blasting cut coal depends on:

(a) Thickness of the seam.

(b) Position of the cut.

(c) Presence of hard dirt bands.

(d) Type of parting with the roof.

(e) Cleat of the coal.

(f) Method of working.

Normally, seams up to 4 ft in thickness can be blasted by one row of holes; in thicker seams, two rows may be required.

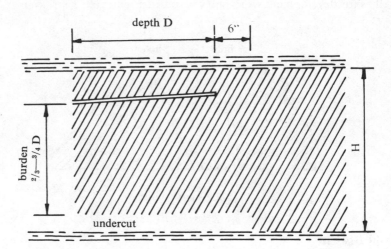

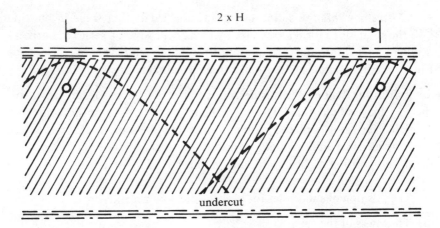

Fig. 73. Blasting a longwall face.

In seams containing hard bands of rock, additional holes may be necessary; these are placed near the band to fracture it without unduly shattering the coal.

Generally, two methods of working can be considered: (a) the longwall face, and (b) room-and-pillar. Examples of shothole placements are shown in Figures 73—75. Figure 73 shows the placement of holes on a longwall face in a 4 ft thick seam, undercut to a depth of 5 ft. In thicker seams, two rows of holes would probably be necessary. Figure 74 shows how holes are placed in a room-and-pillar face 12 ft wide in a 5 ft thick seam. In Figure 75, the placement of holes is shown in a room-and-pillar face 12 ft wide, in a 5 ft thick seam with a 4 inches "dirt" band. The centre hole is fired first, followed by the two top or two bottom holes.

Delay Firing

The advantages of delay firing in coal mining operations are now accepted. In addition to the practical advantages of better fragmentation, reduced ground vibration, and low explosive consumption, delay firing reduces the hazards from ejected fragments, from dust, and from falls of ground.

The exploitation of the advantages of delay firing in coal has been restricted due to the possibility of firing into a break or firing an open suspended shot in the presence of inflammable gas. The development of safer explosives has permitted the introduction of delay firing in coal.

Safe Charging Procedures

In breaking cut coal, several important points of procedure should be noted, whether for longwall or room-and-pillar workings, and independent of the position of the cut.[4]

1. The cut (kerf) should be advantageously located.

2. Cuttings or duff should be removed from the kerf before firing.

3. Shot holes should not extend to or beyond the rear end of the kerf, or behind the line of rib. Ample clearance should be allowed.

4. Holes should be drilled as nearly horizontal as possible.

5. Holes should be thoroughly cleaned out before charging.

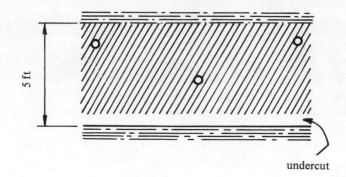

5 ft

undercut

Fig. 74. Blasting in a room-and-pillar face.

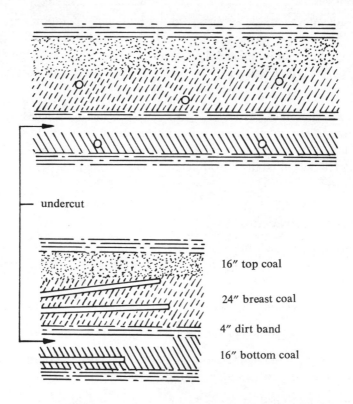

undercut

16″ top coal

24″ breast coal

4″ dirt band

16″ bottom coal

Fig. 75. Blasting in a room-and-pillar face with a dirt band.

6. Holes (and the face generally) should be closely examined for cracks or breaks. If noted, they should be reported to the supervisor.

7. Holes showing cracks or breaks should not be charged for firing.

8. Care should be taken not to overload nor undercharge a hole.

9. Non-incendive short-delay copper shell type detonators should be used, along with a suitable type of permissible explosive charge.

10. Holes should not be charged if the methane concentration at the face exceeds one per cent.

11. Charges should be continuously loaded to the bottom of the hole with cartridges of the same type.

12. Allowable charge weights per hole should not be exceeded.

13. Adequate non-combustible stemming material should be used.

Substitutes for Explosives

In order to avoid the potential hazard of methane/coal dust explosions, several non-explosive methods of breaking coal have been devised.

1. **High pressure gas cartridges:**[5] These depend upon the release of gas pressure from a steel tube capsule vented suddenly into a borehole.

Typical of these are Cardox and Airdox, as described in Chapter 10. Permissible models of Cardox blasting devices are listed periodically by MESA together with specified requirements for their employment in coal mines.[2]

2. **High pressure water jet drilling:**[6] This method should have wide acceptance in breaking coal, in that the coal is wetted and kept cool, dust is allayed at the source or formed as a suspension in water, and extraneous hazards are avoided (see Chapter 10).

Pulsed Infusion Shotfiring

The coal blasting technique called **pulsed infusion shotfiring** was developed in 1953. It consists of firing an explosive charge in a borehole filled with water under pressure, the water having been introduced through an infusion tube, which also seals the hole. Water pressures of the order of 400 lb/in^2 are used. The energy from the explosive is used more efficiently in pulsed infusion shotfiring than when blasting in the conventional manner.

A gelatinous explosive of high water-resistance is required.[8] There are two methods of applying pulsed infusion shotfiring:

1. Flanking holes applicable to longwall faces (both solid and undercut), and to room-and-pillar workings; and

2. Longhole method — shotholes up to 150 ft long are drilled parallel to the face with a burden of 3 ft. The charge consists of a series of high velocity gelatine cartridges. Applications are in room-and-pillar workings, with holes bored from one end of a pillar to the other; and in longwall working, holes being bored parallel to the face.

References

1. U.S. Code of Federal Regulations, 30 CFR, Chap. 1, Subchapter C, Part 15, July 1, 1975 (Schedule 1H).

2. J. RIBOVICH, R. W. WATSON and J. J. SEMAN, "Active List of Permissible Explosives and Blasting Devices Approved Before December 31, 1975", *Mining Enforcement and Safety Administration, I.R. 1046, 1976*.

3. *Blaster's Handbook*, 6th ed. Canadian Industries Ltd., 1966.

4. *Blaster's Handbook*, 15th ed. E. I. du Pont de Nemours & Co (Inc.), 1967.

5. R. E. GREENHAM and H. STAFFORD, "Recent American Developments No. 2. Shotfiring and Its Alternatives in the United States of America", *Trans. I.M.E.*, Vol. 113, p. 814 (1953).

6. D. A. SUMMERS, "Water Jet Coal Mining Related to the Mining Environment", *University of Missouri-Rolla*, Conference on the Underground Mining Environment, October 1971.

7. "Water Jet Technology" (Exotech Inc., 1969).

8. R. HASLAM, S. H. DAVIDSON and J. HANCOCK, "Development of a Combined Blasting/Water Infusion Technique for Coal Breaking", *Trans. I.M.E.*, Vol. 114, p. 87 (1954).

CHAPTER 21

Explosives in Agriculture

Introduction

Explosives have a very definite place in agriculture, just as they have in mining, quarrying, and construction work. As a source of power, they enable the time-consuming and heavy work of clearing the land of trees, tree-stumps, and boulders to be done in a fraction of the time and at less cost than by hand methods. Swamp drainage, stream diversion, subsoiling, and tree planting are further operations in which explosives may be employed successfully.

In agricultural work, the quantity of explosives required for any particular operation may vary considerably according to local conditions; therefore the quantities mentioned should be regarded only as a guide, and more accurate determinations obtained by actual trials under prevailing conditions.[1]

The chief uses of explosives in agriculture may be set out as follows:

Removal of Stumps and Trees

1. **General principles:** The method employed in the removal of stumps varies according to the age, size, and type of stump, nature of the soil, ground conditions, character of root system, and equipment available. All these factors must be taken into consideration when determining the best and most economical method of removing stumps by blasting.

The nature and condition of the soil is of considerable importance in stump blasting. The more resistance the soil offers to the force of the explosion, the greater will be the force exerted against the stump. Loose, sandy soil, which is dry, is far less efficient in this regard than firm, heavy, wet soil. Light sandy soils provide a ready escape for the explosive gases; therefore, it is always advisable to blast when the soil is wet. As a rule, charges should be placed deeper in light soils than in heavy soils.

Depending on the type of tree with its associated root system, the methods used to remove stumps may vary somewhat. However, the main requirement is to blast a conical cavity which encloses the main root system of the stump. Some trees have heavy tap roots, some lateral spreading roots, others have both. Methods of placing the explosive charge must be varied to suit the particular stump. In addition, green stumps have a large mass of fine tendril-like roots, and these normally require a heavier charge to dislodge them than dead stumps on which the tendril roots have decayed.

2. **Preparation of charges:** As a basis on which to carry out trial blasts, about one pound of ready-mixed AN/FO should be used for each ft of stump diameter measured 1 ft above the ground surface where dead stumps are involved. With green stumps, it may be necessary to increase the charging ratio slightly. With standing trees, charges may have to be increased two or three times. The presence of sandy rather than clay soil also makes an increase in charge necessary (of the order of 50 per cent).

The use of small plastic bags makes the placement of AN/FO more convenient. The required charge is placed in the plastic bag together with one-half of a cartridge of dynamite primed with a plain detonator and safety fuse, or an electric detonator. The plastic bag also serves to protect the soluble AN from any water which may be present in the hole.

3. **Placing the charges:** A little experience on the part of the operator will soon allow rapid determination of the correct location for the charges. However, the following details will serve as a guide. Generally speaking, it is felt that the holes to accommodate the charges should be dug rather than drilled, as the charges can then be located as a whole rather than poured into place, and contamination with soil, which may cause loss of sensitivity, is avoided. A narrow spade is satisfactory as it makes digging between roots easier.

Small stumps up to about 2 ft in diameter with lateral spreading roots should normally be removed with one charge placed under the centre of the stump or as close to that position as possible. This is illustrated in Figure 76.

However, it often happens that the roots are heavier on one side than the other, and in such cases the charge would be located towards the side with the heavier roots, as in Figure 77.

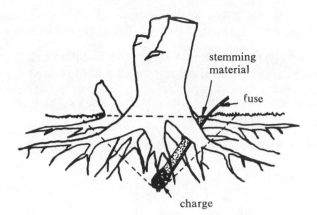

Fig. 76. Typical charge placement for small tree stumps.

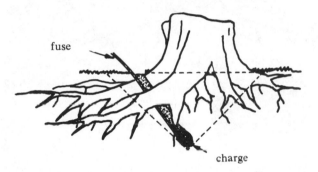

Fig. 77. Typical charge placement for tree stump with asymmetrical root pattern.

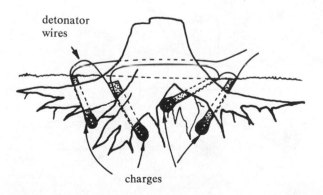

Fig. 78. Typical charge placement for large tree stumps with large lateral root system.

Larger stumps with lateral spreading roots (Fig. 78) may require two or more charges, these normally being located under the heaviest roots.

Stumps with large tap roots may be removed by placing a large charge against the tap root at a depth of not less than 2 ft as illustrated in Figure 79.

After the charge is in position, the soil or clay should be firmly replaced to provide maximum confinement.

Hollow stumps are dealt with as suggested in Figure 80.

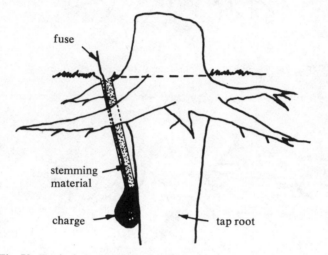

Fig. 79. Typical charge placement for tree stumps with large tap root.

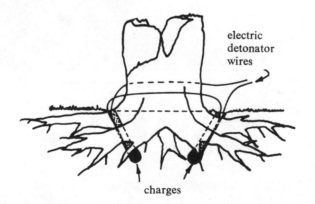

Fig. 80. Typical charge placement for hollow tree stumps.

4. **Methods of firing:** Where one charge only is to be fired under each stump, safety fuse may be used. However, if two or more charges under the stump are involved, it is essential that they detonate simultaneously, and electric detonators or DC must be used. For electric firing, the EB caps are connected in series and fired with an exploder; while, to use DC, lines from the primer in each of the charges are connected to a main DC line by clove hitch joints. The DC main line is then initiated with a plain detonator and safety fuse. These methods are described in more detail in Chapters 7, 8.

Breaking Boulders

Blasting is the quickest and usually the most economical method of breaking up boulders or rock ledges and reducing them to a size that can be handled more easily.

Three methods can be employed.

1. **Mudcapping** is used when no rock drilling equipment is available; it consists of placing a charge of dynamite, primed with safety fuse and detonator, in close contact with the rock and covering it completely with a liberal quantity of stiff damp clay, as illustrated in Figure 81.

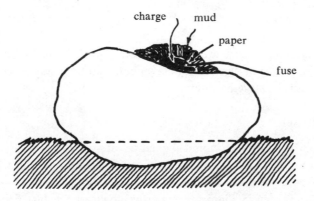

Fig. 81. Mudcapping a boulder.

The charge should be placed at a point where it will perform most effectively, and this is usually on a flat or concave surface on the rock. It is important that the explosive is pressed down in close contact with the rock surface and all air excluded from the plaster; therefore it is often advantageous to put a thin layer of wet clay on the rock and bed the explosive in it.

Charges may vary from 8 to 16 oz/yd³ of rock, according to the type of rock to be broken.

2. **Snakeholing:** Where a boulder is almost buried beneath the surface, this method is used to lift it out of the ground and break it at the same time. It consists of making a hole under the boulder and placing a charge immediately below it, as in Figure 82. The charge should be calculated on a basis similar to the previous method.

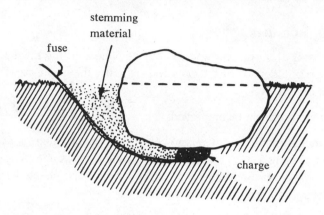

Fig. 82. Snakeholing a boulder.

3. **Blockholing** consists of drilling a hole 6 in deep into a boulder, and charging it with 1 oz of dynamite. Several holes, strategically placed, may be required with very large boulders.

Log Splitting

Probably the most economical method of splitting logs up to 3 ft in diameter is to use a "Log-splitting Gun", various types of which are on the market. Essentially, it consists of a solid mild steel cylinder 14—18 inches long, by 1¹/₂ inches in diameter, tapering down in the last 6 inches to a diameter of slightly over ⁵/₈ inches. At the tapered end is a ⁵/₈ inch hole, 7—8 inches long, into which, at the closed end, is drilled a touch hole for the safety fuse. The hole is filled to within ¹/₂ inch of the top with Blasting Powder (approximately 1¹/₂ oz); the tapered end of the gun is then driven into the end of the log for an inch or two, and the charge of powder touched off with the safety fuse. The log is split by the gases developed from the ignition of the Blasting Powder.

For logs over 3 ft in diameter and exceeding 5 ft in length, it may be necessary to bore one or more holes into the side of the log and charge them with dynamite, AN/FO, or Blasting Powder. It is important that the depth of the holes should be almost two-thirds of the diameter in order to locate the charge near the centre of the log. Logs 3 ft in diameter and 5 ft long can be split in four sections with one 4 oz charge, whereas a log 4 ft in diameter and 12 ft in length will require a total charge of 12—16 oz spread over three equally spaced holes. The holes must be charged and stemmed in the usual manner.

In hardwoods, the above method involves considerable labor in boring the holes unless a power drill is available.

Ditching and Draining

Ditches can be excavated in certain types of soil by blasting, but the practicability of this can only be determined by trial. As a general rule, it can be stated that blasting is most effective in moist or wet loam and in clay (providing it is not too sticky), whereas it is usually ineffective in sand and gravel, and in hard-packed dry earth.

There are two distinct methods of blasting ditches: by the propagation method, and by the electrical method. The former requires an explosive which is very sensitive to the propagation of a detonation wave, and can be used in wet loamy soils, while the latter can be employed in most soils except sand without the use of special explosives. Straight dynamite 50% (Ditching dynamite) is preferred for both methods.

1. **Propagation method:** In the propagation method, only one charge of explosive is primed with a detonator and safety fuse. When this is detonated, the detonation wave is transmitted through the moist soil to the adjoining charges along the line of ditch. The action takes place with great rapidity, so that the whole row of charges explodes practically simultaneously. The maximum distance through which propagation will take place between individual charges is dependent upon the characteristics of the soil and the size of the charges. In mud, a single row of 8 oz charges, placed in vertical holes 18 inches deep, and spaced 18 inches apart, should provide a ditch approximately 3 ft deep, and 8 ft wide at the top. For wider ditches of the same depth, two or more rows of holes would be necessary. Where greater depth is desired, the holes must be made deeper and the charge quantity increased. For ditches over 6 ft in depth, larger charges, spaced at consider-

able distance apart, and fired electrically, will prove more economical. Holes for the charges which are to be propagated should be made with a pinch bar or stick to a depth such that the top of the charge is not over 1 ft below the surface. The charges should be stemmed, either with water or soil, to obtain maximum efficiency. There is little hope of success for this method in a heavy, clayey soil, or in sand.

2. **Electrical method:** The electrical method of ditching must always be used when soil is insufficiently wet for the charges to propagate, and when an explosive sensitive to propagation is unavailable. With this method, each charge must be primed with an electric detonator; when charging is completed the detonators are connected in a series circuit and fired simultaneously by an exploder (see Fig. 18).

The depth of the holes, number of rows, weight of charge, and spacing will depend on the size and nature of ditch required, and should be determined by trial. For small ditches 3 ft in depth, holes $1^1/_2$ ft deep, spaced at 2 ft intervals, and charged with 8 oz of dynamite should prove satisfactory.

The length of ditch that can be blasted at one time is limited by the capacity of the exploder.

Tree Planting

The use of explosives for tree planting has a two-fold purpose. Not only is the hole made, but the ground is loosened over an area considerably in excess of that which would have been affected by the spade. As in all soil blasting, the work must be done when the soil, especially clay soil, is dry. A vertical hole bored to a depth of 3 ft and charged with 4—8 oz of explosive, well confined, should prove satisfactory.

Subsoiling

An impervious subsoil or hard pan may be treated with explosives with satisfactory results, but apart from this the practice commends itself to all agriculturists on account of its effect upon the soil at a depth many feet below that to which the plough can reach. Best results will be obtained when the ground is in a dry and friable state. When the ground is wet, the action of the explosive may cause compression of the surrounding soil, thereby aggravating its condition instead off effecting improvements.

The operation consists of boring holes to a depth of 2—3 ft at intervals of 12—20 ft according to the hardness of the soil, and charging each hole with about 4 oz of explosive primed with detonator and safety fuse. The stemming material should be tightly tamped to ensure the effective confinement of the charge. As in most agricultural blasting work, it is advisable to make test shots in order to determine the correct weight of charge, depth of hole, and spacing.

The procedure in detail is as follows:

Rows of holes equivalent to the number of operators are first prepared. These are then loaded and tamped, sufficient fuse being allotted to each hole to ensure the projection of several inches above the ground. When ready to fire, the operators, commencing at one end, and keeping in line, proceed along their respective rows lighting the fuses. In this way the operators are well in advance of the explosions, and in safety (Fig. 83).

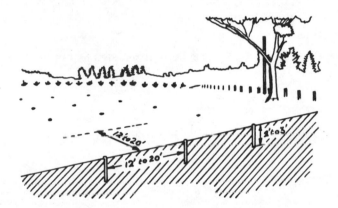

Fig. 83. Subsoiling for orchard development.

Digging Pole Holes

For loosening stiff clay, the surface soil is first removed and a hole punched to the desired depth with a driving bar. A light charge is then placed, stemmed and fired.

In surface rock across which fences or telephone lines or power lines are to be constructed, it is frequently necessary to blast the hole to receive the post or pole. The most difficult task is in drilling a hole to a little lower than the desired depth of the pole hole; especially where drilling equipment is unavailable. Such a hole may be charged with dynamite and well stemmed before firing with cap and fuse. The main objective is to loosen the rock for subsequent removal of the fragments by hand shovel. For deep holes, short holes are drilled and fired in stages, the rock particles being cleaned out after each stage of firing.

A new development is the Phillips PTR 101 Pole Hole Explosive which is a safe-to-handle pre-cast unit that can be fitted with a cap and set on the ground over the hole site. It acts like a simple form of shaped charge. It is more expensive than dynamite but saves considerable time and effort.

Note: The authorities in Canada and in some states of the U.S. may require that licences be obtained before mixing or using ammonium nitrate or any explosives for blasting. This should be discussed with the local Police Officer; or Inspector of Explosives or Inspector of Mines.

References

1. *Explosives Bulletin* No. 5, Nobel (Australasia) Pty Ltd. October 1948.

CHAPTER 22

Laws Relating to Explosives

General

Persons handling and using explosive materials should realize at all times that they are not inherently absolutely safe; they are only statistically so, within broad limits of confidence. Therefore, the utmost caution should attend their use. In simple terms, explosives are dangerous but most explosives accidents could be prevented.

In handling, some types (e.g., blasting agents) are much less sensitive than dynamites. However, all are designed ideally to be relatively insensitive to handle and store when manufactured to close tolerances and inspection procedures; but to explode violently under controlled stimuli when put to use, with the release of tremendous units of energy within a mere fraction of a second.

Departures from this ideal design in the shape of substandard manufacturing performance; or with improper handling or prolonged storage under adverse conditions; or by inappropriate methods of use, can lead to various levels of calamity.

There are no specific restrictive laws governing the manufacture of explosives. With this class of product, the need for safety is paramount; manufacturers strive to preserve their reputations, and their urge to produce a reliable safe product, as members of the IME, transcends the profit motive. Any departure from a firstclass standard of manufacture, inspection, distribution and service is likely to lead to expensive damage claims.

With explosives, unless mixed and incorporated properly, the mixed explosive may only be as safe as the properties of each separate ingredient.

However, some manufacturers offer no warranties and others issue disclaimers under certain conditions of use. Ethically, these policies cannot be recognized as avenues of escape from just penalties due to shortfalls from firstclass standards of manufactured products.

The employment of explosives as an engineering tool has been referred to in Chapter 1; it has made a significant contribution to our standard of living by increasing the productivity of the rock breaking process, and by reducing its arduous nature.

However, if used incautiously, or with malicious intent, damage to property and personal injury (including loss of life) can readily result, either to the user, or to innocent bystanders. [Suffice it to say that any malicious use of explosives that results in personal injury or loss of human life should be regarded as a bestial offence. The perpetrator should surely forfeit all constitutional and other rights to be rated as a human being. It represents a degree of social irresponsibility that can only be expiated and provide a realistic deterrent, in the interests of society, by capital punishment].

For all the above reasons, it has become necessary to formulate laws and regulations prescribing acceptable methods of manufacture, transportation, storage, distribution and use. These will be dealt with serially, as under.

United States

1. **Manufacture:** In statutory terms,there are two aspects of manufacture of explosive materials (other than sporting ammunition) in the United States: (a) Manufacture of the complete range of explosives and accessories by established firms at their various factory sites, and (b) the mixing of blasting agent components at or near the blast site by mining companies, quarry operators, contractors, or explosives manufacturers.

In either case, a licence is necessary under the provisions of Title XI, Regulation of Explosives, see below.[1]

In order to manufacture "permissible" explosives, such products must be tested and approved by MESA, under Schedule 1-H.[2]

2. **Storage:** This procedure normally occurs at several different stages between manufacture and use, e.g.:

(1) At the factory, awaiting distribution.

(2) At regional points, to give more effective local distribution.

(3) At dealers' depots.

(4) At mixing depots, in the case of contract mixing services, for blasting agents.

(5) At mines and quarries, and on contractors' premises.

(6) For underground mines, at suitable storage points.

(7) At docksides, for ocean and river transport.

Such storage depots are referred to as "magazines". Various regulations, referred to below, prescribe conditions of location, construction, maintenance, operation and surveillance. All must be licensed and/or are subject to periodic inspection.

Various stages of transportation must necessarily occur between storage points.

Explosives magazines must not be located within certain prescribed distances of inhabited buildings, passenger railway stations, public highways, and of other magazines, according to the maximum weight of high explosives stored (see Appendix A).

Similarly, a facility for storing ammonium nitrate or blasting agents (or for mixing these components) must not be located within a limiting distance, as a receptor unit, of an explosive magazine (described as a donor) in case of sympathetic detonation from the latter (see Appendix B).

3. **Transportation:** Regulations covering the packaging and transportation of explosive materials by land are under the control of the Chief Inspector, Bureau of Explosives, which is a unit of the U.S. Department of Transportation (DOT). Such regulations appear as CFR Title 49.[3]

Similar regulations for sea transport and dockside storage are administered by the U.S. Coast Guard,[4] and by the Civil Aeronautical Board, for air transport.

The DOT classifies explosive materials for transportation as under:

Class A: Explosives possessing detonating or otherwise maximum hazard; such as dynamite, NG, picric acid, lead azide, fulminate of mercury, black powder, blasting caps, and detonating primers.

Class B: Explosives possessing flammable hazard, such as propellant explosives.

Class C: Oxidizing material (blasting agents, ammonium nitrate, etc.).

The DOT regulations do not apply to military explosives, or to other federal agencies in the performance of official duties, or to fireworks, or to transport on private property (such as on farms, mines or quarries). However, mix trucks and pump trucks when traversing public roads between mixing plant and mines are subject to DOT (as well as to local and state) regulations.

4. Use: Commercial explosives are used in a variety of ways, such as in mines (surface and underground), quarries, construction jobs, farms etc. Its use in mines is governed federally by MESA[5, 6, 7, 8] from the health and safety standpoint, and by state regulations, in many cases.

Various state and municipal authorities regulate explosive usage within the 50 states. There is no uniformity among these jurisdictions and there is insufficient space here to list their separate requirements. The particular authority holding jurisdiction within any city, county or state may be the Chief Inspector of Mines, the Chief of Police or the Chief of the Fire Department. Persons interested in using explosives should approach any of these officials (in the city concerned, or in the capital city of the state) for details and permits to use and blast.

The National Fire Protection Association has issued a Code for the manufacture, transportation, storage, and use of explosives and blasting agents.[9] It incorporates the DOT Regulations in its chapter on transportation and the Internal Revenue Service (IRS) Regulations in its chapters on security and storage (see below). The Code is recommended for adoption as city and state regulations to promote safety and security in respect of explosive materials. The NFPA does not itself have any inspectorial or approving authority.

The Occupational Safety and Health Regulations of the Department of Labor (in sub-part H) also apply to the use of commercial explosives.

The Institute of Makers of Explosives (IME) issued its Publication No. 3 in 1970.[10] Apart from a special section designed to establish close control over the unauthorized possession and use of explosives (anticipating new Federal laws for this purpose) it appears to follow closely the NFPA Code 495.

Both the above publications[9, 10] have included an appendix designed to guide legislation in state bills with the laudable aim of promoting some reasonable degree of uniformity in the control of explosive materials.

Public Law 91—452 became effective on February 12, 1971, as Title XI, Regulation of Explosives, of the "Organized Crime Control Act of 1970."[1]

Section 1101 of the U.S. Code Title 18 states that "The Congress hereby declares that the purpose of this title is to protect interstate and foreign commerce against interference and interruption by reducing the hazard to persons and property arising from misuse and unsafe or insecure storage of explosive materials."

Regulations under this act were promulgated on January 15 and became effective on February 12, 1971 as Title 26, Chapter 1, Subchapter E Part 181 "Commerce in Explosives" under the aegis of the Department of the Treasury, Internal Revenue Service.

Inter alia, the regulations are designed to cover:

The issue of licences to manufacturers, importers and dealers.

The issue of permits to users.

The conduct of business operations (relating to explosives distribution transactions).

The maintaining of records and reports (relating to explosives distribution transactions).

The location, construction, repair, control and inspection of storage facilities (magazines).

The expeditious reporting of thefts and losses.

Five different categories of magazines are set out, together with detailed specifications for the construction and operation of each.

Canada

The Canadian "Explosives Act", administered by the Department of Energy, Mines and Resources at Ottawa, through the Chief Inspector of Explosives, is a statute enacted in the interests of public safety. It controls the manufacture, authorization, sale, storage, importation and transportation by road of explosives, through a system of licences and permits. The licensing requirement for the manufacture of explosives includes on-site mixing.[11]

All explosives must be submitted for authorization tests prior to manufacture or importation. Regulations specify the requirements for the transportation of explosives by road.[14] Permits are required for road transport of quantities in excess of 2000 kg. The transport of explosives by rail is regulated by the Board of Transport Commissioners in accordance with the "Regulations for the Transportation of Dangerous Commodities by Rail".

Explosives in Canada are divided into seven classes: gunpowder, nitrate mixtures, nitro-compounds, chlorate mixtures, fulminate, ammunition and fireworks, numbered from 1 to 7 in that order.

Regulations specify the details of construction of and the licensing requirements for explosives storage magazines other than for military establishments.[13] A Vendor's Magazine Licence is required for the sale of any quantity of explosives. A User's Magazine Licence is required for the storage, for private use, of 75 kg or 2000 detonators. A licence is not required for the storage of lesser quantities for private use.

Detonators must be stored separately from other explosives. A detonator magazine must be at least 150 ft from a dynamite magazine. Detonating cord must not be stored with detonators but in a dynamite magazine. Safety fuse is not to be stored with detonators.[12] Magazines must be located at minimum prescribed distances from railways, highways, points of public assembly, dwelling houses, shops, churches, and schools, in accordance with the Table of Safe Location Specifications (Appendix C).

Careful records should be kept of all receipts and issues of explosive materials, and of all sales transactions, including the name and address of each purchaser.

Explosives that have deteriorated are especially hazardous and must be carefully destroyed by experienced personnel according to methods laid down.

Permitted explosives only may be used in coal mines; they are tested and approved by MESA or the U. K. Ministry of Energy.

All members of the Royal Canadian Mounted Police are authorized to act as Deputy Inspectors of Explosives.

The actual usage of explosives is regulated by the various provincial authorities. Persons interested should contact these authorities for details.

References

1. Department of the Treasury, Internal Revenue Service, "Organized Crime Control Act of 1970", Title XI — Regulation of Explosives. Public Law 91—452. October 15, 1970.

2. U.S. Code of Federal Regulations, Schedule 1-H, "Explosives and Related Articles, Procedure for Testing for Permissibility and Related Tests". CFR Title 30, Chapter 1, Sub-chapter C, Part 15.

3. U.S. Department of Transportation Regulations. CFR Title 49, Parts 1—49, 71—78, 190—197.

4. U.S. Coast Guard Regulations. CFR Title 33, Parts 6 and 126. CFR Title 46, Sub-chapter N, Parts 146—147.

5. U.S. Code of Federal Regulations, "Mandatory Safety Standards, Underground Coal Mines". CFR Title 30, Chapter 1, Sub-chapter 0, Part 75, Sub-Part N. November 20, 1970.

6. U.S. Code of Federal Regulations, "Health and Safety Standards — Metal and Non-metallic. Open Pit Mines". CFR Title 30, Chapter 1, Part 55 (as amended).

7. U.S. Code of Federal Regulations, "Health and Safety Standards — Metal and Non-metallic. Underground Mines". CFR Title 30, Chapter 1, Part 57 (as amended).

8. U.S. Code of Federal Regulations, "Health and Safety Standards — Sand, Gravel and Stone Mines". CFR Title 30, Chapter 1, Part 56 (as amended).

9. National Fire Protection Association, "Code for the Manufacture, Transportation, Storage and Use of Explosives and Blasting Agents", No. 495, 1973.

10. Institute of Makers of Explosives, "Suggested Code of Regulations for the Manufacture, Transportation, Storage and Use of Explosive Materials", Pub. No. 3. June 1970.

11. Department of Energy, Mines and Resources, Canada. "Explosives Act and Regulations (as amended)" 1970.

12. Department of Energy, Mines and Resources, Canada. "The Storage of Explosives" 1972.

13. Department of Energy, Mines and Resources, Canada. "Standard for Blasting-Explosives Magazines", 1971.

14. Department of Energy, Mines and Resources, Canada. "Trucking Explosives in Canada". 1970.

Appendix

Appendix A

Safe Location Specifications for Storage of Explosives (USA)

(American Table of Distances for Storage of Explosives as revised and approved
by the Institute of Makers of Explosives – Nov. 1971)

Quantity of Explosives		Distances in Feet								
		Inhabited Buildings		Public Highways Class A to D		Passenger Railways – Public Highways with Traffic Volume of more than 3,000 Vehicles/Day		Separation of Magazines	
Pounds Over	Pounds Not Over	Barricaded	Unbarricaded	Barricaded	Unbarricaded	Barricaded	Unbarricaded	Barricaded	Unbarricaded
2	5	70	140	30	60	51	102	6	12
5	10	90	180	35	70	64	128	8	16
10	20	110	220	45	90	81	162	10	20
20	30	125	250	50	100	93	186	11	22
30	40	140	280	55	110	130	206	12	24
40	50	150	300	60	120	110	220	14	28
50	75	170	340	70	140	127	254	15	30
75	100	190	380	75	150	139	278	16	32
100	125	200	400	80	160	150	300	18	36
125	150	215	430	85	170	159	318	19	38
150	200	235	470	95	190	175	350	21	42
200	250	255	510	105	210	189	378	23	46
250	300	270	540	110	220	201	402	24	48
300	400	295	590	120	240	221	442	27	54
400	500	320	640	130	260	238	476	29	58
500	600	340	680	135	270	253	506	31	62
600	700	355	710	145	290	266	532	32	64
700	800	375	750	150	300	278	556	33	66
800	900	390	780	155	310	289	578	35	70
900	1,000	400	800	160	320	300	600	36	72
1,000	1,200	425	850	165	330	318	636	39	78
1,200	1,400	450	900	170	340	336	672	41	82
1,400	1,600	470	940	175	350	351	702	43	86
1,600	1,800	490	980	180	360	366	732	44	88
1,800	2,000	505	1,010	185	370	378	756	45	90
2,000	2,500	545	1,090	190	380	408	816	49	98
2,500	3,000	580	1,160	195	390	432	864	52	104
3,000	4,000	635	1,270	210	420	474	948	58	116
4,000	5,000	685	1,370	225	450	513	1,026	61	122
5,000	6,000	730	1,460	235	470	546	1,092	65	130

Quantity of Explosives		Distances in Feet							
6,000	7,000	770	1,540	245	490	573	1,146	68	136
7,000	8,000	800	1,600	250	500	600	1,200	72	144
8,000	9,000	835	1,670	255	510	624	1,248	75	150
9,000	10,000	865	1,730	260	520	645	1,290	78	156
10,000	12,000	875	1,750	270	540	687	1,374	82	164
12,000	14,000	885	1,770	275	550	723	1,446	87	174
14,000	16,000	900	1,800	280	560	756	1,512	90	180
16,000	18,000	940	1,880	285	570	786	1,572	94	188
18,000	20,000	975	1,950	290	580	813	1,626	98	196
20,000	25,000	1,055	2,000	315	630	876	1,752	105	210
25,000	30,000	1,130	2,000	340	680	933	1,866	112	224
30,000	35,000	1,205	2,000	360	720	981	1,962	119	238
35,000	40,000	1,275	2,000	380	760	1,026	2,000	124	248
40,000	45,000	1,340	2,000	400	800	1,068	2,000	129	258
45,000	50,000	1,400	2,000	420	840	1,104	2,000	135	270
50,000	55,000	1,460	2,000	440	880	1,140	2,000	140	280
55,000	60,000	1,515	2,000	455	910	1,173	2,000	145	290
60,000	65,000	1,565	2,000	470	940	1,206	2,000	150	300
65,000	70,000	1,610	2,000	485	970	1,236	2,000	155	310
70,000	75,000	1,655	2,000	500	1,000	1,263	2,000	160	320
75,000	80,000	1,695	2,000	510	1,020	1,293	2,000	165	330
80,000	85,000	1,730	2,000	520	1,040	1,317	2,000	170	340
85,000	90,000	1,760	2,000	530	1,060	1,344	2,000	175	350
90,000	95,000	1,790	2,000	540	1,080	1,368	2,000	180	360
95,000	100,000	1,815	2,000	545	1,090	1,392	2,000	185	370
100,000	110,000	1,835	2,000	550	1,100	1,437	2,000	195	390
110,000	120,000	1,855	2,000	555	1,110	1,479	2,000	205	410
120,000	130,000	1,875	2,000	560	1,120	1,521	2,000	215	430
130,000	140,000	1,890	2,000	565	1,130	1,557	2,000	225	450
140,000	150,000	1,900	2,000	570	1,140	1,593	2,000	235	470
150,000	160,000	1,935	2,000	580	1,160	1,629	2,000	245	490
160,000	170,000	1,965	2,000	590	1,180	1,662	2,000	255	510
170,000	180,000	1,990	2,000	600	1,200	1,695	2,000	265	530
180,000	190,000	2,010	2,010	605	1,210	1,725	2,000	275	550
190,000	200,000	2,030	2,030	610	1,220	1,755	2,000	285	570
200,000	210,000	2,055	2,055	620	1,240	1,782	2,000	295	590
210,000	230,000	2,100	2,100	635	1,270	1,836	2,000	315	630
230,000	250,000	2,155	2,155	650	1,300	1,890	2,000	335	670
250,000	275,000	2,215	2,215	670	1,340	1,950	2,000	360	720
275,000	300,000	2,275	2,275	690	1,380	2,000	2,000	385	770

Explanatory Notes Essential to the Application of the American Table of Distances for Storage of Explosives

Note 1 — "Explosive materials" means explosives, blasting agents, and detonators.

Note 2 — "Explosives" means any chemical compound, mixture, or device, the primary or common purpose of which is to function by explosion. A list of explosives determined to be within the coverage of "18 U.S.C. Chapter 40, Importation, Manufacture, Distribution and Storage of Explosive Materials" is issued at least annually by the Director of the Bureau of Alcohol, Tobacco, and Firearms of the Department of Treasury.

Note 3 — "Blasting agents" means any material or mixture, consisting of fuel and oxidizer, intended for blasting, not otherwise defined as an explosive: Provided, That the finished product, as mixed for use or shipment, cannot be detonated by means of a number 8 test blasting cap when unconfined.

Note 4 — "Detonator" means any device containing a detonating charge that is used for initiating detonation in an explosive; the term includes, but is not limited to electric blasting caps of instantaneous and delay types, blasting caps for use with safety fuses and detonating-cord delay connectors.

Note 5 — "Magazine" means any building or structure, other than an explosives manufacturing building, used for the permanent storage of explosive materials.

Note 6 — "Natural Barricade" means natural features of the ground, such as hills, or timber of sufficient density that the surrounding exposures which require protection cannot be seen from the magazine when the trees are bare of leaves.

Note 7 — "Artificial Barricade" means an artificial mound or revetted wall of earth of a minimum thickness of three feet.

Note 8 — "Barricaded" means that a building containing explosives is effectually screened from a magazine, building, railway, or highway, either by a natural barricade, or by an artificial barricade of such height that a straight line from the top of any sidewall of the building containing explosives to the eave line of any magazine, or building, or to a point twelve feet above the center of a railway or highway, will pass through such intervening natural or artificial barricade.

Note 9 — "Inhabited Building" means a building regularly occupied in whole or in part as a habitation for human beings, or any church, schoolhouse, railroad station, store, or other structure where people are accustomed to assemble, except any building or structure occupied in connection with the manufacture, transportation, storage or use of explosives.

Note 10 — "Railway" means any steam, electric, or other railroad or railway which carries passengers for hire.

Note 11 — "Highway" means any street or public road. "Public Highways Class A to D" are highways with average traffic volume of 3,000 or less vehicles per day as specified in "American Civil Engineering Practice" (Abbett, Vol. 1, Table 46, Sec. 3-74, 1956 Edition, John Wiley and Sons).

Note 12 — When two or more storage magazines are located on the same property, each magazine must comply with the minimum distances specified from inhabited buildings, railways, and highways, and, in addition, they should be separated from each other by not less than the distances shown for "Separation of Magazines," except that the quantity of explosives contained in cap magazines shall govern in regard to the spacing of said cap magazines from magazines containing other explosives. If any two or more magazines are separated from each other by less than the specified "Separation of Magazines" distances, then such two or more magazines, as a group, must be considered as one magazine, and the total quantity of explosives stored in such group must be treated as if stored in a single magazine located on the site of any magazine of the group, and must comply with the minimum of distances specified from other magazines, inhabited buildings, railways, and highways.

Note 13 — Storage in excess of 300,000 lbs. of explosives in one magazine is generally not required for commercial enterprises; however, IME will provide recommendations for quantities greater than 300,000 lbs. in one magazine upon inquiry.

Note 14 — This Table applies only to the manufacture and permanent storage of commercial explosives. It is not applicable to transportation of explosives or any handling or temporary storage necessary or incident thereto. It is not intended to apply to bombs, projectiles, or other heavily encased explosives.

For transportation purposes, the Department of Transportation in Title 49 Transportation CFR Parts 1-199 subdivides explosives into three classes:

Class A — Maximum Hazard

Class B — Flammable Hazard

Class C — Minimum Hazard

Note 15 — All types of blasting caps in strengths through No. 8 cap should be rated at $1^1/_2$ lbs. of explosives per 1,000 caps. For strengths higher than No. 8 cap, consult the manufacturer.

Note 16 — For quantity and distance purposes, detonating cord of 50 to 60 grains per foot should be calculated as equivalent to 9 lbs. of high explosives per 1,000 feet. Heavier or lighter core loads should be rated proportionately.

Safe Location Spezifications for Storage of Blasting Agent Materials (USA)
(Table of recommended Separation Distances of Ammonium Nitrate and Blasting Agents
from Explosives or Blasting Agents)[1,6]

Donor Weight		Minimum Separation Distance of Acceptor when Barricaded[2] (ft.)		Minimum Thickness of Artificial
Pounds Over	Pounds Not Over	Ammonium Nitrate[3]	Blasting Agent[4]	Barricades[6] (in.)
–	100	3	11	12
100	300	4	14	12
300	600	5	18	12
600	1,000	6	22	12
1,000	1,600	7	25	12
1,600	2,000	8	29	12
2,000	3,000	9	32	15
3,000	4,000	10	36	15
4,000	6,000	11	40	15
6,000	8,000	12	43	20
8,000	10,000	13	47	20
10,000	12,000	14	50	20
12,000	16,000	15	54	25
16,000	20,000	16	58	25
20,000	25,000	18	65	25
25,000	30,000	19	68	30
30,000	35,000	20	72	30
35,000	40,000	21	76	30
40,000	45,000	22	79	35
45,000	50,000	23	83	35
50,000	55,000	24	86	35
55,000	60,000	25	90	35
60,000	70,000	26	94	40
70,000	80,000	28	101	40
80,000	90,000	30	108	40
90,000	100,000	32	115	40
100,000	120,000	34	122	50
120,000	140,000	37	133	50
140,000	160,000	40	144	50
160,000	180,000	44	158	50
180,000	200,000	48	173	50
200,000	220,000	52	187	60
220,000	250,000	56	202	60
250,000	275,000	60	216	60
275,000	300,000	64	230	60

Notes

Note 1 — Recommended separation distances to prevent explosion of ammonium nitrate and ammonium nitrate-based blasting agents by propagation from nearby stores of high explosives or blasting agents referred to in the Table as the "donor." Ammonium nitrate, by itself, is not considered to be a donor when applying this Table. Ammonium nitrate, ammonium nitrate-fuel oil or combinations thereof are acceptors. If stores of ammonium nitrate are located within the sympathetic detonation distance of explosives or blasting agents, one-half the mass of the ammonium nitrate should be included in the mass of the donor.

Note 2 — When the ammonium nitrate and/or blasting agent is not barricaded, the distances shown in the Table shall be multiplied by six. These distances allow for the possibility of high velocity metal fragments from mixers, hoppers, truck bodies, sheet metal structures, metal containers, and the like which may enclose the "donor." Where storage is in bullet-resistant magazines[1] recommended for explosives or where the storage is protected by a bullet-resistant wall, distances and barricade thicknesses in excess of those prescribed in the American Table of Distances are not required.

Note 3 — The distances in the Table apply to ammonium nitrate that passes the insensitivity test prescribed in the definition of ammonium nitrate fertilizer promulgated by the Fertilizer Institute;[2] and ammonium nitrate failing to pass said test shall be stored at separation distances determined by competent persons and approved by the authority having jurisdiction.

Note 4 — These distances apply to nitrocarbonitrates and blasting agents which pass the insensitivity test prescribed in regulations of the U.S. Department of Transportation and the U.S. Department of the Treasury, Bureau of Alcohol, Tobacco and Firearms.

Note 5 — Earth, or sand dikes, or enclosures filled with the prescribed minimum thickness of earth or sand are acceptable artificial barricades. Natural barricades, such as hills or timber of sufficient density that the surrounding exposures which require protection cannot be seen from the "donor" when the trees are bare of leaves, are also acceptable.

Note 6 — For determining the distances to be maintained from inhabited buildings, passenger railways, and public highways, use the Table of Distances for Storage of Explosives in Appendix A of *NFPA 495-1973, Code for the Manufacture, Transportation, Storage, and Use of Explosive Materials*.

[1]For construction of bullet-resistant magazines see Chapter 3 of *NFPA 495-1973, Code for the Manufacture, Transportation, Storage, and Use of Explosive Materials*.

[2]Definition and Test Procedures for Ammonium Nitrate Fertilizer, Fertilizer Institute, November 1964.

Appendix C
Safe Location Specifications for Storage of Explosives (Canada)
(Quantity-Distance Table for Blasting Explosives)
(Table Volume – Distance pour Explosifs Industriels)

Quantity of Blasting Explosives and Related Explosive Accessories *Volume d'explosifs industriels et accessoires explosifs*	Distance from: a) Highway, street and other road accessible to the public. b) Railway. c) Aerodrome and airfield. d) Bank of navigable waterway including canals and recreational waters. e) Park, playground, sportsfield and other open places of resort. f) Work area.	*Distance de:* *a) Grande route, rue et voie publique.* *b) Voie ferrée.* *c) Champ d'aviation et aéroport.* *d) Berge de voie navigable (y compris canaux et lieux de récréation).* *e) Parc, terrain de jeux, stade ou autre lieu public à découvert.* *f) Lieu de travail.*	Distance from: a) Dwelling, living, quarters. b) Store, office and like buildings. c) School, college or university. d) Hospital and rest home. e) Church and chapel. f) Theatre, auditorium and other assembly buildings. g) Factory, workshop, service station and garage. h) Warehouse or storage area for flammable substances in bulk.	*Distance de:* *a) Demeures et habitations.* *b) Magasin, bureau et immeuble du genre.* *c) École, collège ou université.* *d) Hôpital et maison de retraite.* *e) Chapelle et église.* *f) Théâtre, auditorium et autres lieux publics.* *g) Usine, atelier, stationservice et garage.* *h) Entrepôt ou zone de stock de matières inflammables.*	Distance between traversed magazines *Distrance entre dépôts d'explosifs avec remblai ou tertre*
1 Kilogr. *Kilogr.*	2 Metres *Mètres*		3 Metres *Mètres*		4 Metres *Mètres*
50	23		23		9
100	23		32		11
200	26		52		14
250	30		60		15
300	34		68		16
400	41		82		18
500	47		94		19
600	53		105		20
800	65		130		23
1,000	75		150		24
2,000	120		240		30
4,000	175		350		38
5,000	190		380		41
6,000	200		400		44
7,000	210		420		46
10,000	240		480		52
20,000	300		600		66
25,000	320		640		70
30,000	340		680		74
40,000	380		760		82
50,000	410		820		88
100,000	525		1,050		110
150,000	588		1,175		128

Appendix D
Weights of Various Rock Materials

Material	Density	Solid			Broken	
		Lb./ Cu. Ft.	Cu. Ft./ Ton	Tons/ Cu. Yd.	Lb./ Cu. Ft.	Cu. Ft./ Ton
Andesite	2.4–2.8	160	12.5	2.16	114.3	17.5
Anorthosite	2.6–2.9	170	11.8	2.29	121.4	16.5
Basalt	2.7–3.2	185	10.8	2.50	132.1	15.1
Brucite	2.3–2.4	145	13.8	1.96	103.6	19.3
Coal-Bituminous	1.2–1.5	85	23.5	1.15	60.7	32.9
Dolomite	2.8–2.9	180	11.1	2.43	128.6	15.6
Diabase	2.8–3.1	185	10.8	2.50	132.1	15.1
Diorite	2.7–3.0	180	11.1	2.43	128.6	15.6
Gabbro	2.9–3.0	185	10.8	2.50	132.1	15.1
Granite	2.5–2.8	165	12.1	2.23	117.9	17.0
Gypsum	2.3–3.3	175	11.4	2.37	125.0	16.0
Hematite	4.5–5.3	305	6.6	4.09	217.9	9.2
Limestone	2.4–2.9	165	12.1	2.23	117.9	17.0
Magnetite	5.0–5.2	320	6.3	4.29	228.6	8.8
Mica Schist	2.5–2.9	170	11.8	2.29	121.4	16.5
Norite	2.7–3.0	180	11.1	2.43	128.6	15.6
Nepheline Syenite	2.5–2.7	160	12.5	2.16	114.3	17.5
Peridotite	3.1–3.3	200	10.0	2.7	142.9	14.0
Porphyry	2.5–2.6	160	12.5	2.16	114.3	17.5
Quartz	2.65	165	12.1	2.23	117.9	17.0
Quartzite	2.4–2.8	160	12.5	2.16	114.3	17.5
Rock Salt	2.1–2.6	145	13.8	1.96	103.6	19.3
Rhyolite	2.4–2.6	155	12.9	2.09	110.7	18.1
Sandstone	2.0–2.8	150	13.3	2.03	107.1	18.7
Shale	2.4–2.8	160	12.5	2.16	114.3	17.5
Siderite	3.0–3.9	215	9.3	2.90	153.6	13.0
Slate	2.5–2.8	165	12.1	2.23	117.9	17.0
Talc	2.6–2.8	170	11.8	2.29	121.4	16.5
Trap Rock	2.6–3.0	175	11.4	2.37	125.0	16.0
Tuff	2.0–2.6	145	13.8	1.96	103.6	19.3

Appendix E
Explosives Loading Table
(Pounds of Explosives per ft of Blasthole)

Hole Diameter (inches)	Specific Gravity												
	0.8	0.9	1.0	1.10	1.15	1.20	1.25	1.30	1.35	1.40	1.45	1.50	1.60
1	0.27	0.31	0.34	0.37	0.39	0.41	0.43	0.44	0.46	0.48	0.49	0.51	0.54
1¼	0.42	0.48	0.53	0.58	0.61	0.64	0.67	0.69	0.72	0.74	0.77	0.80	0.85
1½	0.61	0.69	0.77	0.84	0.88	0.92	0.96	0.99	1.03	1.07	1.11	1.15	1.22
1¾	0.83	0.94	1.04	1.14	1.20	1.25	1.30	1.35	1.41	1.46	1.51	1.56	1.66
2	1.09	1.22	1.36	1.50	1.57	1.63	1.70	1.77	1.84	1.90	1.97	2.04	2.18
2½	1.70	1.92	2.13	2.34	2.45	2.56	2.66	2.77	2.87	2.98	3.09	3.20	3.41
3	2.45	2.75	3.06	3.37	3.52	3.67	3.83	3.98	4.14	4.28	4.44	4.59	4.90
3½	3.33	3.75	4.17	4.58	4.80	5.00	5.21	5.41	5.63	5.83	6.05	6.25	6.66
4	4.35	4.90	5.44	5.98	6.26	6.53	6.81	7.07	7.36	7.62	7.90	8.16	8.70
4½	5.51	6.20	6.89	7.58	7.93	8.27	8.62	8.96	9.31	9.65	10.00	10.34	11.02
5	6.80	7.65	8.50	9.35	9.79	10.20	10.64	11.05	11.49	11.90	12.34	12.75	13.60
5½	8.23	9.26	10.29	11.32	11.84	12.35	12.88	13.38	13.91	14.41	14.94	15.44	16.46
6	9.79	11.02	12.24	13.46	14.10	14.69	15.32	15.91	16.55	17.14	17.77	18.36	19.58
6½	11.50	12.93	14.37	15.81	16.54	17.24	17.98	18.68	19.42	20.12	20.86	21.56	22.99
7	13.33	14.99	16.66	18.33	19.19	19.99	20.86	21.66	22.52	23.32	24.19	24.99	26.66
7½	15.30	17.22	19.13	21.04	22.03	22.96	23.94	24.87	25.86	26.78	27.77	28.70	30.61
8	17.41	19.58	21.76	23.94	25.06	26.11	27.24	28.29	29.42	30.46	31.60	32.64	34.82
8½	19.66	22.11	24.57	27.03	28.29	29.48	30.75	31.94	33.21	34.40	35.67	36.86	39.31
9	22.03	24.79	27.54	30.29	31.72	33.05	34.48	35.80	37.23	38.56	39.99	41.31	44.06
9½	24.55	72.62	30.69	33.76	35.34	36.83	38.36	39.90	41.43	42.97	44.50	46.04	49.10
10	27.20	30.60	34.00	37.40	39.16	40.80	42.56	44.20	45.97	47.60	49.37	51.00	54.40
10½	29.99	33.74	37.49	41.24	43.17	44.99	46.86	48.74	50.61	52.49	54.36	55.24	59.98
11	32.91	37.03	41.14	45.25	47.38	49.37	51.43	53.48	55.54	57.60	59.65	61.71	65.82
11½	35.98	40.47	44.97	49.47	51.72	53.96	56.21	58.46	60.71	62.96	65.21	67.46	71.95
12	39.17	44.06	48.96	53.86	56.39	58.75	61.29	63.65	66.20	68.54	71.10	73.44	78.34

Glossary of Terms

AN — Ammonium nitrate

AN/FO — A blasting agent consisting of AN and fuel oil.

AN slurry — *See* Slurry

Aluminized — *See* Metallized.

Back-break — Ground broken beyond the limits of the last or outer row of holes.

Binary explosive — A 2-component explosive based on safe-to-handle compounds such as hydrazine or nitromethane, shipped separately and united at the site to form a high-energy explosive.

Blasthole — *See* Borehole.

Blasting agent — Any material or mixture consisting of a fuel (combustible) and oxidizer, intended for blasting, not otherwise classified as an explosive and in which none of the ingredients is classified as an explosive — provided that the finished product, as mixed and packaged for use or shipment, cannot be detonated by a No. 8 test blasting cap when confined.

Blasting pattern — *See* Pattern.

Blasting Powder — A low explosive consisting of potassium or sodium nitrate, charcoal, and sulphur.

Blistering — *See* Mud-capping.

Blockholing — A method of breaking a boulder with explosives placed in a small borehole.

Booster — An explosive of special character, generally used in small quantities to improve the performance of another explosive, the latter constituting the major portion of the charge. When it is armed with a detonator, it becomes a primer.

Borehole — A hole drilled into rock to accommodate an explosives charge for blasting (breaking) the rock.

Bottom priming — A method in which the primer is placed near the bottom of the charge or hole.

Brisance — The ability of an explosive to break (or shatter) rock by shock or impact, as distinct from gas pressure.

Bulled hole — A shothole which has been bulled or chambered by exploding a light charge of high explosive in the bottom; a hole with a chamber at the bottom to accommodate a large quantity of explosive.

Burden — The volume of rock which lies within the zone of influence of a charge of explosive; the volume of rock to be broken by any one hole or charge.

Burden distance — The distance between the main body of a charge and the nearest free face.

Butt — The remaining part of a hole after the charge has been fired; a hole which was not fully blown out; a bootleg. It is a likely place to find an unexploded charge and it is therefore dangerous to try to extend it by further drilling.

Cap — Detonator.

Capped fuse — Safety fuse to which a plain detonator has been crimped.

Cartridge — A preformed unit of high explosive wrapped to a predetermined diameter and length; a plug; stick of dynamite; a soft plastic stick of AN/FO or slurry.

Chapman-Jouget plane — The C-J plane may be described simply as that "point" along a confined cylindrical column of explosives at which the detonation shock wave reaction is complete; and from which point onwards a permanent steady-state reaction continues.

Collar priming — A method in which the primer is placed near the top, or at the collar end of the charge.

Cooling salt — Either sodium chloride or sodium carbonate incorporated in a high explosive to reduce the heat of the explosion as in permitted (permissible) explosives. A flame-depressant.

Cordeau detonant — *See* Detonating cord.

Coupling, degree of — The extent to which an explosives charge is in direct contact with the rock wall of the hole.

Critical mass — Combustion, or burning, is a term usually employed to describe a reaction between a fuel and atmospheric air. This typically occurs when a small quantity of explosives burns.

When a considerable quantity of explosives is burning, an adiabatic reaction may develop between the ingredients, reinforced by heat transferred from the gaseous products of the reaction.

The threshold quantity of explosives so required to produce this second class of reaction is known as the "critical mass", which has a characteristic value for each explosive substance. In quantities above the critical mass, therefore, quiescently burning explosives may suddenly spontaneously explode.

This propensity is particularly important to be aware of when destroying deteriorated or unwanted explosives by burning (see Chapter 11).

Cut — An artificial opening made in a face to provide a free face for blasting. It may be made mechanically as in coal mining or by explosives. *See also* Kerf.

Demolition — The breaking up of artificial (man-made) structures by blasting.

Detonating cord (DC) — A strong flexible cord containing a core of detonating explosive, used primarily for initiating a series of charges. It explodes practically instantaneously throughout its length, when initiated with a detonator.

Detonating wave — The shock wave set up when a detonator is ignited.

Detonator — A cap or capsule of sensitive explosive material used to initiate a charge of high explosives.

"Dirt" — A band of stone or shale in a coal face.

Fragmentation — The extent to which rock is broken up into small pieces by (primary) blasting.

Gelatinous — Plastic. A property of those types of high explosive that can be pressed into different shapes, e.g., blasting gelatine, gelatine dynamite, etc. Gelatinous explosives are much more readily and efficiently placed or tamped in a hole without including pockets of air.

Gunpowder — *See* Blasting Powder.

HE — High explosive (dynamite).

Hole — *See* Borehole.

Ignite, to — The act of lighting; to set alight. A safety fuse is ignited by a flame. A plain detonator is ignited by safety fuse. An electric detonator is ignited by a current that fuses a fusible bridgehead embedded in a flashing mixture. Black powder is ignited by safety fuse.

Igniter cord (IC) — A cord filled with thermite for readily igniting a multiplicity of safety fuses in sequence.

Initiate, to — The act of detonating a high explosive by means of a detonator or by detonating cord.

Jumbo — A highly integrated, mobile drilling rig on which the drills are mounted on booms manoeuvred (usually) by hydraulic controls.

Kerf — A cut made in a coal face by a hand pick or by a mechanical coal cutter to provide a free face for blasting.

Kibble — Bucket in which broken ore or rock is hoisted from a winze.

Leg wires — The wires connecting the electrodes of an electric detonator; the wires which are permanently attached to the electric detonator.

Lifter — The bottom holes of a round pattern for a drift or tunnel heading.

Metallized — In respect of blasting agents, sensitized or boostered with metal powders or granules (usually aluminium or ferrosilicon), to yield more energy.

Misfire — A charge or part of a charge which for one of any number of reasons has not exploded. They are usually difficult and dangerous to resolve. Misfires must be treated with respect.

Mud-capping — A method of breaking a boulder with explosives without drilling a hole. The charge is set in a depression on the boulder and covered with a thick plaster of wet clay to confine it.

Muck — The broken ore or rock resulting from the firing of one or more charges.

NG — Nitroglycerine.

Overbreak control — A method of firing perimeter holes in such a way as to avoid intensive fracturing of a wall with the aim of preserving a regular outline (cushion blasting; pre-splitting).

Pattern — A dimensioned plan of holes to be drilled for blasting a face.

Permitted (Permissible) — A term signifying that a particular explosive has been approved as suitable for use in coal mines.

Plastering — See Mud-capping.

Plug — See Cartridge.

Powder — A general term designating any commercial explosive, irrespective of type.

Prills — Cellular sub-globular particles of AN formed by spraying concentrated AN solution against a stream of air.

Primer — A cartridge of HE (or a prefabricated precast "booster") incorporating a detonating device. This is the key element of a charge of explosives.

Pull, pulled — The quantity of rock excavated in a unit of advance in a heading, as by drilling and blasting a round.

Ring drilling — A method of drilling a fan-shaped pattern of holes in a single plane from one drill rig position.

Round — The series of blastholes required to produce a unit of advance in a development heading or other face.

Scram drift — An underground drift in which ore is transported (transferred) by drag scraper.

Sequential firing — A system of firing in which the holes with least resistance are detonated progressively, thus reducing the burden on each subsequent hole in turn.

Shothole — *See* Borehole.

Slurry — An aqueous solution of AN sensitized with a combustible fuel (and thickened with a gelling agent at the point of charging).

SN — Sodium nitrate.

Spacing — The linear distance between collars of shotholes bored approximately parallel to a free face.

Stemming — Sandy clay material prepared and wrapped in cartridge form and used for sealing a blast-hole after the charge has been placed.

Stick — "Stick of powder"; plug, cartridge.

Stope — The chamber or excavation from which ore is extracted underground.

Strength — The explosive strength of unit weight (or volume) of a high explosive when compared with that of Blasting Gelatine in a ballistic mortar. Although compared with Blasting Gelatine it is sometimes designated in percentage of nitroglycerine (%NG). This latter designation is not a true measure of its strength.

Strike — The direction along which a lode or vein intersects the surface.

Tamping — The act of charging or tamping a charge into a hole, with the aid of a tamping stick. Sometimes used loosely for "stemming" *(q.v.)*.

TNT	— Trinitrotoluene.
Top priming	— *See* Collar priming.
Toe	— A remmant of rock left unbroken at the foot of a quarry face by an unsatisfactory blast (see Fig. 43).
USSS	— United States Standard Sieve.
VOD	— Velocity of detonation, a measure of the rate at which the detonating wave travels through an explosive charge; the speed of detonation of a particular explosive.

INDEX

The letter G signifies that the item also appears in the Glossary

Important Publications for
Civil and Mining Engineers